Narrow Tilting Vehicles

Mechanism, Dynamics, and Control

Synthesis Lectures on Advances in Automotive Technology

Editor
Amir Khajepour, *University of Waterloo*

The automotive industry has entered a transformational period that will see an unprecedented evolution in the technological capabilities of vehicles. Significant advances in new manufacturing techniques, low-cost sensors, high processing power, and ubiquitous real-time access to information mean that vehicles are rapidly changing and growing in complexity. These new technologies—including the inevitable evolution toward autonomous vehicles—will ultimately deliver substantial benefits to drivers, passengers, and the environment. Synthesis Lectures on Advances in Automotive Technology Series is intended to introduce such new transformational technologies in the automotive industry to its readers.

Narrow Tilting Vehicles: Mechanism, Dynamics, and Control
Chen Tang and Amir Khajepour
2019

Dynamic Stability and Control of Tripped and Untripped Vehicle Rollover
Zhilin Jin, Bin Li, and Jungxuan Li
2019

Real-Time Road Profile Identification and Monitoring: Theory and Application
Yechen Qin, Hong Wang, Yanjun Huang, and Xiaolin Tang
2018

Noise and Torsional Vibration Analysis of Hybrid Vehicles
Xiaolin Tang, Yanjun Huang, Hong Wang, and Yechen Qin
2018

Smart Charging and Anti-Idling Systems
Yanjun Huang, Soheil Mohagheghi Fard, Milad Khazraee, Hong Wang, and Amir Khajepour
2018

Design and Avanced Robust Chassis Dynamics Control for X-by-Wire Unmanned
Ground Vehicle
Jun Ni, Jibin Hu, and Changle Xiang
2018

Electrification of Heavy-Duty Construction Vehicles
Hong Wang, Yanjun Huang, Amir Khajepour, and Chuan Hu
2017

Vehicle Suspension System Technology and Design
Avesta Goodarzi and Amir Khajepour
2017

Narrow Tilting Vehicles: Mechanism, Dynamics, and Control
Chen Tang and Amir Khajepour

ISBN: 978-3-031-00373-8   paperback
ISBN: 978-3-031-01501-4   ebook
ISBN: 978-3-031-00006-5   hardcover

DOI 10.1007/978-3-031-01501-4

A Publication in the Springer series
SYNTHESIS LECTURES ON AVANCES IN AUTOMOTIVE TECHNOLOGY

Lecture #7
Series Editor: Amir Khajepour, *University of Waterloo*
Series ISSN
Print 2576-8107   Electronic 2576-8131

Narrow Tilting Vehicles

Mechanism, Dynamics, and Control

Chen Tang and Amir Khajepour
University of Waterloo

*SYNTHESIS LECTURES ON ADVANCES IN AUTOMOTIVE TECHNOLOGY
#7*

ABSTRACT

To resolve the urban transportation challenges like congestion, parking, fuel consumption, and pollution, narrow urban vehicles which are small in footprint and light in their gross weight are proposed. Apart from the narrow cabin design, these vehicles are featured by their active tilting system, which automatically tilts the cabin like a motorcycle during the cornering for comfort and safety improvements. Such vehicles have been manufactured and utilized in city commuter programs. However, there is no book that systematically discusses the mechanism, dynamics, and control of narrow tilting vehicles (NTVs).

In this book, motivations for building NTVs and various tilting mechanisms designs are reviewed, followed by the study of their dynamics. Finally, control algorithms designed to fully utilize the potential of tilting mechanisms in narrow vehicles are discussed. Special attention is paid to an efficient use of the control energy for rollover mitigation, which greatly enhance the stability of NTVs with optimized operational costs.

KEYWORDS

narrow tilting vehicle, urban transportation, tilting mechanism, vehicle modeling, envelope control, vehicle control, rollover mitigation

Contents

Preface

This book presents the developments of narrow tilting vehicles (NTVs) from mechanism designs, dynamical analysis, and tilting control perspectives. Motivations for building such narrow vehicles to be used in urban transportation are reviewed in Chapter 1, followed by the necessity for active tilting control to enhance the vehicle roll stability. Tilting mechanisms, which are novel structures to allow the vehicle tilting motions, are classified and reviewed in Chapter 2. Challenges in tilting mechanism designs are discussed, with the latest endeavour from a mechatronics point-of-view presented. The dynamical model of NTVs is developed in Chapter 3. A novel re-configurable modelling approach is introduced to handle various wheel and actuator configurations in NTV applications. Roll stability criterion are derived in detail, and suspension designs with tilting stability in mind are also presented. Chapter 4 proposes an energy-efficient tilting control scheme, and concludes the book with various implementations of the proposed envelope-based tilting control.

Chen Tang and Amir Khajepour
June 2019

Acknowledgments

This book would not have been possible without the help of many people. We are particularly grateful to Avesta Goodarzi, Mansour Ataei, Iman Fadakar, Amir Soltani, and Ling He for their inspiring discussions, and our family for their patience, love, and support.

The authors are grateful for the financial support of the Natural Sciences and Engineering Research Council of Canada (NSERC) and Ontario Research Fund in conducting this research.

Both authors are also thankful to Morgan & Claypool Publishers for providing the opportunity for this book, along with their consistent encouragement and support throughout this project.

Chen Tang and Amir Khajepour
June 2019

CHAPTER 1

Urban Vehicles and Narrow Tilting Vehicles

1.1 ISSUES IN URBAN TRANSPORTATION

As the number of vehicles in big cities has dramatically increased in recent years, people have had to face urban transportation issues like congestion, parking, and pollution. The average number of occupants per vehicle, as investigated by the U.S. Department of Transportation, is found to be 1.58 [1]. Conventional four-wheeled vehicles, which are designed to accommodate four to six passengers comfortably as well as providing sufficient space for their cargo, are considered unnecessarily large for their average passenger load in normal city driving [2]. The surplus sizing of the vehicle takes more space when driving on the road as well as parking, and the extra weight comes along with an inevitable increase in fuel consumption.

On the other hand, two-wheeled transportation tools, such as bicycles, mopeds, and motorcycles, are considered extremely space and fuel efficient [3] for the personal mobility. However, passenger safety and weather protection designs of such two-wheelers are not as good as their four-wheeled counterparts. Conventional automobiles also perform significantly better for their payload-carrying capability as well as the on-board electronics for enhanced active safety. Apart from that, drivers of two-wheelers also need to learn how to balance the vehicle under various conditions as it is inherently unstable at low speeds, which also limits its public acceptance.

To address the above-mentioned issues in urban transportation systems, concepts of urban vehicles are proposed [2, 4] for the niche market of personal mobility by combining the benefits of conventional cars with those of two-wheelers. They are designed with recent advancements in automotive mechatronics but in a narrow footprint to seat limited passengers. The reduced size saves on production cost and utilizes existing road infrastructure (e.g., lane and parking space) more efficiently. The reduction in the gross mass also helps to improve fuel efficiency and cut pollution, which gives it a competitive advantage in emerging car-sharing programs.

1.2 NARROW URBAN VEHICLE AS A SOLUTION

The previous section demonstrated the need for compact urban vehicles which carry limited passengers in pursuit of significantly improved fuel efficiency while not compromising on the safety and comfort offered by standard four-wheeled cars [3]. Due to the unstable nature of two-wheelers, three- and four-wheeled narrow vehicle designs are usually adopted to address urban transportation challenges. Such vehicles are designed to provide just the right space to seat one or two people in tandem for a daily commute, and the narrow footprint enables them to be operated on reduced-size lanes and parked in compact spaces. Benefiting from the enclosed cabin, a weatherproof interior and a crash-worthy body structure could be designed for a more comfortable and safer ride. The concept of such narrow urban vehicles is visualized in Figure 1.1.

Figure 1.1: Concepts of narrow urban vehicles on the road. (Image generated using CARLA simulator [5].)

The design of such vehicle with a small footprint seems to be the cure for many urban transportation issues, but it has its own problem—such vehicles are less stable against rollover compared with conventional cars. A good field-of-view for human drivers should be maintained which means the riding height of a narrow urban vehicle should be similar to a conventional street car, as illustrated in Figure 1.2. With its track width reduced almost by half, a narrow vehicle becomes significantly less stable. According to static stability factor (SSF), which was adopted by the new car assessment program (NCAP) as one of the ratings for vehicle rollover resistance from 2001–2003 [6], the rollover danger of narrow vehicles almost doubled compared with regular-width automobiles.

Lateral centrifugal force is the major cause for un-tripped rollovers, and high speeds in tight cornering could generate risky centrifugal forces to endanger vehicle stability. By consider-

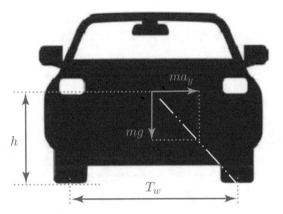 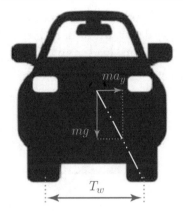

Figure 1.2: Vehicle static rollover and static stability factor.

ing the geometrical aspects of a car (i.e., track-width T_w, and the height of its center-of-gravity h), a rough estimation of the acceleration threshold (a_y^*) to trigger a lift-off of the inner wheel from ground could be determined from Eq. (1.1):

$$a_y^* \approx SSF \cdot g \tag{1.1}$$

with

$$SSF \triangleq \frac{T_w}{2h}.$$

Generally speaking, a "narrow" vehicle is defined as any vehicle with its SSF smaller than 0.6 [2]. Vehicles with high SSF are considered safer against rollover danger since their tire forces saturate before fatal centrifugal forces can be developed. This inherently prevents vehicle rollover as a fail-safe system. Unfortunately, this does not hold for narrow vehicles. Such vehicles could reach their tipping limit at lateral accelerations less than 0.7 g while the skidding limit is around 0.6–0.8 g [2].

1.3   TOWARD NARROW TILTING VEHICLES

The solution is again inspired by two-wheelers. If such tall and narrow vehicles can actively lean into the curve like motorcycles during cornering, their roll stability can be enormously enhanced, with passenger comfort and driving experience improved as well. Such tilting systems are thus crucial in narrow vehicle designs. Apart from this, the tilting system should be designed to function without direct human driver interventions, so that any driver who can operate a conventional vehicle can quickly adapt to a narrow vehicle counterpart without worrying about balancing the roll motion. This calls for the need for an active tilting system with automatic control to guarantee vehicle stability and safety under the harshest driving scenarios [7]. Narrow urban vehicles equipped with such tilting systems are known as narrow tilting vehicles (NTVs).

Figure 1.3 summarizes the motivation for NTVs based on previous discussions. The arrows in grey denote the turning direction of the vehicle, while the red arrows illustrate the distribution of the normal load as an indicator of vehicle roll stability.

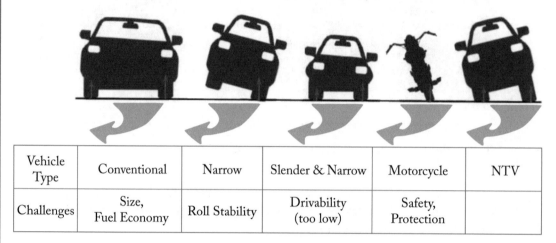

Vehicle Type	Conventional	Narrow	Slender & Narrow	Motorcycle	NTV
Challenges	Size, Fuel Economy	Roll Stability	Drivability (too low)	Safety, Protection	

Figure 1.3: Motivation for NTVs.

Various tilting system designs have been proposed [8–11], and several NTV prototypes have been built by the automotive industry since 1950.

The Ford Gyron was a gyroscopically stabilized two-wheeled vehicle with retractable wheels pods. The gyroscope, which weighted 180 lb, stabilized the vehicle and allowed it to travel at a lateral acceleration up to 1 g in cornering [2]. However, the extra weight of the gyroscope, at the same time, became a limitation of the vehicle.

Another attempt was made by General Motors in the 1970s, when a three-wheeler called the Lean Machine was fabricated. It was composed of a tilting cabin and a non-tilting engine assembly. With a motorcycle front-end for steering, the body could rotate up to 55° with respect to the non-tilting part [2]. The tilting motion was operated by human drivers through a foot pedal, which required extra control efforts.

In 1997, Mercedes-Benz unveiled their NTV concept called F-300 Life-jet [12]. The vehicle was characterized by a two-wheeled front axle which can lean up to 30° and a single rear wheel with the trailing arm suspension. An active tilting control system was equipped, but the car was still designed to be almost as wide as a conventional car with a front track width of 1,560 mm.

Carver One and the CLEVER (compact low emission vehicle for urban transport) were models developed in year 2002 and 2003, respectively. Both of them had a single steerable front wheel mounted on the tilting cabin and equipped with a non-tilting axle with two wheels at the back. Their configurations were similar to that of the Lean Machine, with a tilting capability up to 45° using hydraulic actuators under an automatic control system named direct tilt control (DTC) [13].

Smera was a four-wheeled tilting vehicle developed by Lumeneo. This electric vehicle was regulated as a car in Europe and had a maximum speed of 128.7 km/h with a range of 145 km on a single charge [14]. The on-board tilt control system was able to tilt the vehicle by a maximum angle of 20°.

One of the recent endeavors in NTV was the Toyota i-Road [15] released in 2013. The three-wheeled electric vehicle was equipped with the active tiling system to balance the vehicle automatically. The car was designed to travel at a top speed of 45 km/h with a track width of 850 mm.

A comparison of several NTV configurations is summarized in Table 1.1.

Table 1.1: Comparison of tilting vehicle configurations

Model	Wheel Configuration	Steerable Wheels	Driven Wheels	Tilting Parts
F-300	2F1R[3]	Front	Rear	Full
Carver	1F2R[1]	Front	Rear	Partial
Smera	2F2R[2]	Front	Rear	Full
i-Road	2F1R[1]	Rear	Front	Full

[1] Tadpole conguration: 2 wheels at the front and 1 wheel at rear.
[2] Delta conguration: 1 wheel at the front and 2 wheels at rear axle.
[3] Conventional vehicle conguration: 2 wheels at both front and rear.

It should be mentioned that NTVs are still in their early stage of the development, with manufacturers suggesting different configurations, as shown in Table 1.1. Various numbers of wheel modules, location of steer-able and driveable wheels, as well as full/partial vehicle tilting schemes differentiate the techniques for NTV applications. Chapter 2 starts with a discussion on various tilting mechanisms, while different configurations of wheels and actuators will be explored in the modeling (Chapter 3) and control (Chapter 4) sections in rest of the book.

CHAPTER 2

Tilting Mechanisms and Actuators

This chapter reviews mechanisms and actuators to actively tilt narrow urban vehicles. The mechanisms are classified by their capability to either partially or fully tilt the vehicle. Various implementations are reviewed, followed by a summary of the challenges in tilting system designs. A novel design which combines the mechanism and actuators into suspension struts is proposed for NTV applications. Space saving of the compact design as well as its capability to adjust the riding height are examined.

2.1 MECHANISMS FOR PARTIAL TILTING VEHICLES

For partial tilting systems, only the body shell tilts, while the chassis still acts as a fixed base like those in conventional cars. Such mechanisms have been widely used in trains to tilt their heavy payload [16], and several tilting cars like GM Lean Machine, Carver One, and the CLEVER [17–20] adopt such systems in their design.

Since only the cabin is involved in tilting, effectiveness of the system for load balancing is limited, and a large tilting angle is required for rollover mitigation during harsh maneuvers. The tilting and non-tilting parts are connected by a mechanism which allows their relative rotation. Conventional automotive tires can be used on the non-tilting chassis due to their small roll angles, while motorcycle tires are usually seen on the tilting parts to fully utilize the camber thrust and reduce the tire wear due to excessive camber angles.

The schematic of a partial tilting vehicle is illustrated in Figure 2.1a. The tilting cylinders are installed between the non-tilting axis and the tilting cabin [8, 17]. Activation of these cylinders causes a tilting action of the passenger compartment, as shown in Figure 2.1b. According to [21], the driver's torque applied to the steering wheel is measured and used to drive the hydraulic pumps. With the hydraulic oil pressurized into one cylinder and withdrawn from the counteracting cylinder, the cabin can be tilted with regard to the chassis even on bumpy road, as shown in Figure 2.1c.

2.2 MECHANISMS FOR FULL TILTING VEHICLES

Considering NTV's limited space for passengers and cargo, tilting the cabin only is considered less effective for rollover mitigation due to the distribution of its sprung and unsprung masses.

Figure 2.1: Partial tilting systems in different modes: (a) upright mode, (b) tilting mode, and (c) going over a bump during tilting.

It is thus natural to explore full vehicle tilting schemes, which not only tilts the passenger cabin, but the chassis and wheel assemblies as well. The major role of tilting actuators is to generate the relative motion, while the tilting mechanism enforces the synchronized motion between the chassis, the cabin, and wheels. Two implementation examples are detailed below.

One approach is to enforce the motion by parallelogram mechanisms in vehicle roll plane, as the schematic shows in Figure 2.2a. Such a design has been adopted in Mercedes-Benz F-300 as well as recent research prototypes [9, 22, 23].

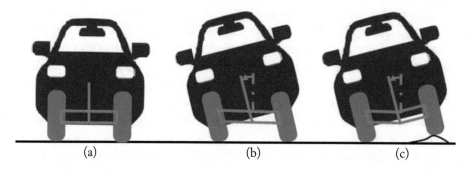

Figure 2.2: Parallelogram tilting mechanisms in different modes: (a) upright mode, (b) tilting mode, and (c) going over a bump during tilting.

A detailed introduction of such mechanical structure is presented in [23]. The whole leaning system is extended from a classical four-bar mechanism, with its body frame, chassis, and wheel assemblies forming the primary linkages, as illustrated in Figure 2.2a. Lengths of the arms are carefully designed to guarantee that the wheels incline with the cabin. The angle between the control arm and the body frame can be altered by using hydraulic or electric actuators, which generates chassis tilting as well as wheel cambers, as shown in Figure 2.2b. NTVs equipped such parallelogram mechanisms in upright mode, tilting mode, and going over a bump during tilting are illustrated, respectively, in Figure 2.2.

The other widely adopted approach is to enforce wheel vertical movements which then generates coupled roll motions via suspension kinematics. The schematic is illustrated in Figure 2.3a, and the implementation using a swing arm mechanism proposed by Edelmann et al. [22] is shown in Figure 2.3b. More examples of such a mechanism can be seen in [10, 14].

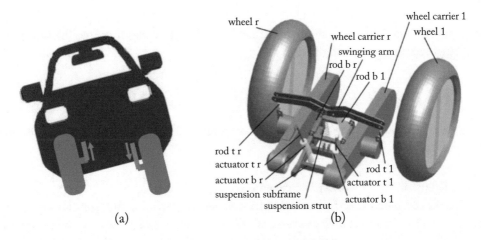

(a)   (b)

Figure 2.3: Full tilting system based on suspension kinematics: (a) schematic, and (b) implementation with a swing arm mechanism [22] (image used with permission).

The idea stems from the fact that if wheel assemblies on both sides can move with regard to the body frame in opposite directions as shown in Figure 2.4b, the vehicle can then actively lean as a whole piece by regulating the suspension motions. Various suspension mechanisms could be utilized as long as enough strokes are allowed by the mechanical designs [24]. Demonstrations of such suspension tilting mechanisms in various modes are illustrated in Figure 2.4.

For the swing arm implementation shown in Figure 2.3b, the suspension sub-frame and wheel carriers are connected to the chassis by revolute joints. Actuators are attached in between to generate the desired tilting motion. The swing arm, along with the rods (br, bl, tr, tl), creates a coupled motion between the wheel carriers on both sides. When one of the wheel carriers goes up, the swing arm mechanism will cause the other wheel carrier to move downward and thus generate the tilting motion. However, when both wheel carriers move uniformly, the suspension sub-frame will rotate relative to the chassis, and the suspension strut can thus absorb road excitations.

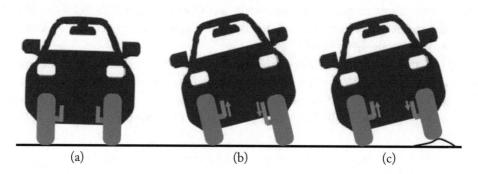

(a)   (b)   (c)

Figure 2.4: Suspension tilting systems in different modes: (a) upright mode, (b) tilting mode, and (c) going over a bump during tilting.

2.3   CHALLENGES IN TILTING MECHANISM DESIGN

After a review of various tilting mechanisms, common challenges in tilting system designs are summarized below.

Effective Tilting for Vehicle Safety Improvements
As mentioned, compared with full-vehicle tilting schemes, partial-tilting solutions [21, 25, 26] are less effective for the roll stability control of NTVs. Only a small portion of the whole vehicle mass is involved in tilting to balance the load, which limits its potential for the emergent and harsh rollover mitigation. However, if the motivation for adopting the tilting system is riding comfort enhancement or driving fun improvement, partial-tilting is still a cost-effective solution due to its relative simple structure and the similarity with conventional vehicles in chassis designs.

System Packaging and Modularity
Parallelogram tilting mechanisms adopted in [9, 11, 14, 23, 26] provide a directly coupled motion between chassis and cabin, and thus generate the full vehicle tilting. However, mechanical linkages for motion synchronization introduce additional complexities to NTVs. Apart from the extra weights added by the system, the connection rods also occupy the cabin space, and impose body design restrictions.

The modularity of the suspension system is also sacrificed by such mechanical designs. Much effort has been devoted to develop urban vehicles in a modularized manner [27–29] to promote the re-usability of their subsystems. The broad adoption of X-by-wire technologies (e.g., steer-by-wire, drive-by-wire, brake-by-wire) removes the steering rod and driving shafts which used to mechanically connect the wheel modules on both sides. Similar technology is expected to simplify the tilting mechanisms for NTVs.

Manufacturing and Operational Costs

The tilting system as an auxiliary active safety system is not expected to significantly increase the manufacturing cost of urban vehicles otherwise their public acceptance might be affected. Existing solutions with exclusive tilting mechanisms and actuators could be optimized to achieve a more cost-effective solution.

Apart from a low cost for production, a reduced operational cost is also desired. With energy concerns in mind, activations of the tilting control should be optimized through control algorithms to be discussed in Chapter 4. From a design perspective, well-designed tilting systems for the NTVs should provide relatively high resistance to rollover in passive mode without sacrificing the riding comfort.

A recent endeavour to address the above-mentioned challenges is presented in [10, 24]. The proposed hydraulic-based integrated suspension tilting system (ISTS) can achieve the same tilting functionality as existing linkage-based tilting mechanisms with less space but more functionalities. The reduced packaging size, as well as the system gross weight savings, can bring many benefits to NTVs. The system is also featured by its enhanced roll stiffness without using an anti-roll bar. All these efforts help to improve the fuel economy, safety, as well as dynamical performances of urban vehicles as a small and lightweight carrier. A brief introduction of the system is presented in the next section.

2.4   INTEGRATED SUSPENSION TILTING SYSTEM

The most significant feature distinguishing the proposed system is the adoption of hydraulic connections to replace the mechanical synchronizer in suspension tilting systems. A direct benefit is the reduction in system weight and complexity. Much space can be saved for passengers and cargo, and hydraulic pipelines can be positioned more easily in the cabin and chassis compared with the mechanical linkages. To make the system even more compact while achieving multi-functionalities, the traditional coil spring and shock absorber are replaced by a hydro-pneumatic suspension, which is a proven technology for the auto industry, as seen in the Citron's Hydractive systems [30, 31].

The schematic of the proposed system on a half-car model is shown in Figure 2.5. The double-acting hydraulic cylinders (4) with accumulators (5) attached to their lower chambers serve as the spring component in conventional suspensions. Orifices, as well as the connecting pipelines, provide the damping sources similar to shock absorbers. The cross inter-connection of the hydraulic cylinders provides a preferable roll stiffness without introducing auxiliary anti-roll bars [10]. Much space and weight can be saved thanks to the simplified structure. The enhanced roll stiffness also helps to reduce the energy consumption due to the less frequent control interventions. Finally, to actively tilt the vehicle, hydraulic pump modules (2) are connected to the upper chambers. By pumping fluid into the system under the coordination of the controller

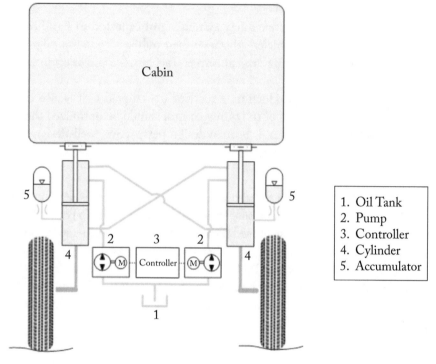

Figure 2.5: Overview of the integrated suspension tilting system (ISTS).

(3), which serves as the software implementation of mechanical constraints, synchronized wheel motions for tilting can be achieved.

As discussed, the synchronization of wheel motions which used to be addressed by mechanical linkages now has a software solution: designing a controller to properly coordinate pump actuation and generating the desired motion. The adoption of separate pump modules also brings an extra degree of freedom: apart from tilting, the system can actively adjust the vehicle riding height. If pistons on both sides travel in an asynchronous mode (same distance but in reverse directions), the tilting motion of the full vehicle can be achieved, as shown in Figure 2.6a. The tilting angle as well as tilting speed is determined by cylinder motions, which can then be regulated via flow rate in hydraulic pumps to offer a responsive feel of tilting. If the pistons move in a synchronous mode (same direction and same magnitude), vehicle height adjustment can be achieved, as shown in Figure 2.6b. Detailed derivations are presented in [10, 32].

Such height adjustment functionality has already been seen on buses [33] to lower their floor for a "kneeling" position. Entering and exiting the car could be much easier especially for the elderly population. On conventional vehicles, such height adjustment has been used to improve the handling and aerodynamics, as well as to increase ground clearance for harsh

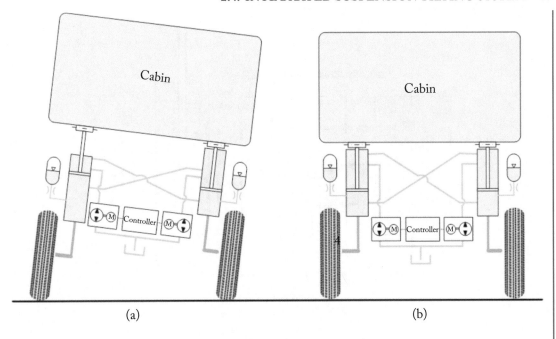

Figure 2.6: ISTS in (a) tilting mode and (b) height adjustment mode.

environments [34]. All these functionalities could be adopted for NTV applications with the proposed system.

Although the idea of an integrated suspension tilting system is presented, the detailed design of the actuation and sensing modules are open for implementations [10]. An example is illustrated in Figure 2.7. The major components are: a fixed displacement pump which can rotate in either direction; a solenoid valve which can disconnect the pump from hydraulic pipelines and eliminates the need for a motor brake; a flow meter to measure the volume of the hydraulic fluid being pumped into the cylinder; and a displacement sensor to monitor the strut movement.

A typical workflow of the ISTS can be described as follows. When a tilting or height adjustment requirement is determined by the vehicle controller, the solenoid valve is first energized, and the fluid will be pumped into the cylinders to drive the piston to the desired position [10]. The process is monitored by the on-board sensing system. Once the desired strut position is achieved, the solenoid valve will be closed. In that state, cylinders and accumulators, along with the connection pipelines, form the hydraulic interconnected suspension (HIS) [31] for passive roll stiffness enhancements.

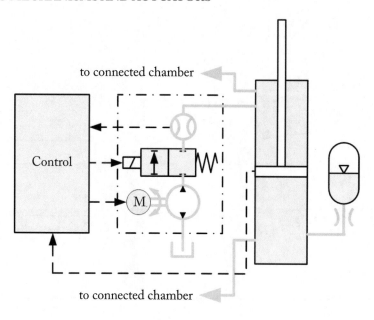

to connected chamber

to connected chamber

Figure 2.7: Actuator, sensor, and controller for ISTS.

2.5   CONCLUSIONS

In this chapter, various mechanisms to generate the tilting motion which are crucial to NTV designs are reviewed. Schematics as well as implementation examples are presented for widely adopted tilting mechanisms. Challenges of tilting system designs are summarized as their effectiveness against rollover, compactness for deployment, and system costs. An ISTS is then introduced, with its motivation, system architecture, and working principles being reviewed as a trial to address such challenges.

CHAPTER 3

Tilting Vehicle Dynamics

This chapter develops mathematical models to capture the dynamical behavior of NTVs. Based on a classical modeling process for conventional automobiles, an integrated handling and roll dynamics for NTV are derived. Topics closely related to NTVs such as derivation of the rollover index, re-configurable actuators, wheel configurations, and suspension kinematics on stability are discussed based on the derived model. The vehicle model and rollover index will be utilized in the controller development in Chapter 4.

3.1 TIRE FORCES AND LATERAL DYNAMICS

For modeling the lateral dynamics of a NTV, one major challenge comes from its flexible wheel configurations. Unlike four-wheeled conventional cars, NTVs can have both three-wheeled and four-wheeled configurations. Among those three-wheelers, there is an extra design flexibility to place the centered wheel either at the front axle (a.k.a. delta configuration), or at the rear axle (a.k.a. tadpole configuration).

To address this in the vehicle modeling process, a general wheel-configuration model in the lateral plane is considered, as shown in Figure 3.1. Six-wheel modules are considered, which are denoted as fl, fc, fr, rl, rc, rr, respectively. By enabling lateral and longitudinal tire forces on corners that match the target vehicle configuration while disabling the rest, all configurations of NTVs can be covered with the proposed generic model. It should also be noted that for any centered wheels, its track width T_{wj} is considered zero.

The lateral and yaw dynamics, considering the net lateral forces F_{yCG} and yaw moment M_{zCG}, can be written as

$$
\dot{v} = \frac{1}{m}F_{yCG} - ur
$$
$$
\dot{r} = \frac{1}{I_z}M_{CG},
$$

(3.1)

where m and I_z are the mass and yaw inertia of the vehicle; and u, v, and r denote vehicle longitudinal speed, lateral speed, and yaw rate, respectively.

Generalized forces F_{yCG} and M_{zCG} on the right-hand-side of Eq. (3.1) are dependent on the tire forces produced at each corner as well as actual wheel configurations. Indexing axles with $i \in \{f, r\}$ (for front and rear axles), and wheels on each axle using $j \in \{l, r, c\}$ (for left, right, and centered wheels), generalized forces and moments applied at the center-of-gravity (CoG)

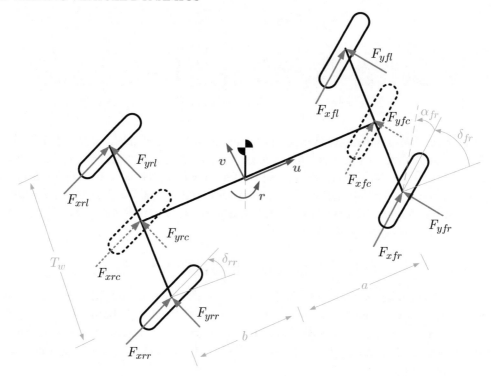

Figure 3.1: General double-track vehicle handling model for an NTV.

can be determined from individual tire force components as

$$F_{yCG} = \sum_{i,j} \left(F_{xij} \sin\left(\delta_{ij}\right) + F_{yij} \cos\left(\delta_{ij}\right) \right)$$

$$M_{zCG} = \sum_{i,j} M_{zij} = \sum_{i,j} \left(\begin{array}{c} \frac{T_{wj}}{2}\left(F_{xij}\cos\left(\delta_{ij}\right) - F_{yij}\sin\left(\delta_{ij}\right)\right) \\ +a_i\left(F_{xij}\sin\left(\delta_{ij}\right) + F_{yij}\cos\left(\delta_{ij}\right)\right) \end{array} \right),$$

(3.2)

where F_{xij} and F_{yij} are longitudinal and lateral forces generated at wheel corner ij; δ_{ij} denotes the steering angle applied on each wheel; and T_{wj} and a_i are generalized track width and axle to CoG distance, respectively. Geometrical parameters a, b, T_w are illustrated in Figure 3.1.

$$T_{wj} = \begin{cases} +T_w & j = \text{r} \\ 0 & j = \text{c} \\ -T_w & j = \text{l} \end{cases} , \quad a_i = \begin{cases} a & i = \text{f} \\ -b & i = \text{r.} \end{cases}$$

The force produced by pneumatic tires F_{xij}, F_{yij} at corner ij has been studied both theoretically and experimentally [35–37]. In general, longitudinal and lateral tire forces can be written as

$$F_{xij} = f_x \left(\alpha_{ij}, S_{ij}, F_{zij}, \gamma_{ij} \right)$$
$$F_{yij} = f_y \left(\alpha_{ij}, S_{ij}, F_{zij}, \gamma_{ij} \right),$$
(3.3)

where α_{ij}, S_{ij}, F_{zij}, and γ_{ij} represent the side slip angle, the slip ratio, the normal force, and the camber angle of each wheel ij, respectively. It should be mentioned that both conventional automotive tires as well as motorcycles tires could be fitted with the model shown in Eq. (3.3) using different parameter sets.

Motorcycle tires generate lateral force mainly through camber angles γ_{ij} known as camber thrust [38], while F_{yij} from conventional automotive tires are dominated by side-slips angles α_{ij}. Differences in contact patch locations [11, 22] between these two types of tires are ignored in this research. The side slip angles α_{ij} and camber angles γ_{ij} at each wheel ij are calculated as

$$\alpha_{ij} = \delta_{ij} - \frac{v + a_i r}{u}$$
$$\gamma_{ij} = K_{\gamma_{ij}} \phi,$$
(3.4)

where ϕ denotes vehicle roll angle and $K_{\gamma_{ij}}$ is the camber-by-roll coefficient determined by suspension and tilting mechanism designs.

For tractability in real-time control applications, an alternative to tire force expressed in Eq. (3.3) is its linearized form. More specifically, for small slip ratios, the longitudinal force contributed by tire ij is proportional to the driving or braking torque Q_{ij} as

$$F_{xij} = \frac{Q_{ij}}{R_{wij}},$$
(3.5)

where R_{wij} denotes the effective rolling radius of wheel ij.

The nonlinear lateral force F_{yij} in Eq. (3.3) could also be approximated by an affine-linear tire model, which linearizes lateral forces at the current operation points of side-slip and camber angles with the zeroth and first-order terms of the Taylor expansion [39, 40]. Denote the lateral tire force, cornering, and camber stiffness of tire ij at the operating point as $\bar{F}_{yij}$, $\tilde{C}_{\alpha i}$, and $\tilde{C}_{\gamma i}$, respectively; the affine-linear lateral tire force model can be expressed as

$$F_{yij} = \bar{F}_{yij} + \tilde{C}_{\alpha ij} \left(\alpha_{ij} - \bar{\alpha}_{ij} \right) + \tilde{C}_{\gamma ij} \left(\gamma_{ij} - \bar{\gamma}_{ij} \right).$$
(3.6)

To capture tire saturation properties, the maximum longitudinal and lateral tire forces should be dependent on the normal force and the road friction conditions. For combined slip scenarios, the tire force constraints can be formulated using the friction ellipse as

$$\left(\frac{F_{xij}}{F_{xij\,\text{max}}} \right)^2 + \left(\frac{F_{yij}}{F_{yij\,\text{max}}} \right)^2 \leq 1,$$
(3.7)

where

$$F_{xij\,\max} = \mu_x F_{zij}$$
$$F_{yij\,\max} = \mu_y F_{zij}$$

and μ_x, μ_y are the friction coefficients in longitudinal and lateral directions, respectively.

3.2   ROLL DYNAMICS AND ROLLOVER INDEX

Roll dynamics is crucial to the operational safety of any NTVs. This section derives a vehicle roll model considering active tilting actuators. Apart from that, as a measure of the rollover tendency, a rollover index based on lateral load transfer (LTR) is also developed.

A generic vehicle roll model is adopted and illustrated in Figure 3.2. The suspension system is approximated using an equivalent rotational spring (K_ϕ) and damper (C_ϕ) connecting the sprung (m_s) and un-sprung mass (m_u) at vehicle roll center (RC). The lateral acceleration (a_y) is treated as a disturbance to the roll model, and the tilting actuator generates an active control moment (T_x) to regulate vehicle roll motions. The net forces at left- and right-side wheels are written as (F_{zl}, F_{zr}) and (F_{yl}, F_{yr}), respectively, for their vertical and lateral components. Geometrical parameters like the distance between sprung mass CoG to roll center (h_s), CoG height of the un-sprung mass (h_u), roll center height (h_{rc}), and track width (T_w) are illustrated in the figure.

The roll dynamics equation can then be derived as,

$$\ddot{\phi} = \frac{1}{I_x}T_x - \frac{m_s h_s \cos(\phi)}{m I_x}a_y - \frac{K_\phi \phi - m_s g h_s \sin(\phi)}{I_x} - \frac{C_\phi}{I_x}\dot{\phi}, \qquad (3.8)$$

where I_x denotes the roll inertia of the vehicle.

By defining the roll rate and roll angle as the states $\left([X_r] = \begin{bmatrix} \dot{\phi} & \phi \end{bmatrix}^T\right)$, tilting moment as the control input ($U_r = [T_x]$), and maneuver-induced lateral acceleration as disturbances ($W_r = [a_y]$), roll dynamics in Eq. (3.8) under the small angle assumption can be written in the state-space form as

$$\dot{X}_r = A_r X_r + B_r U_r + E_r W_r, \qquad (3.9)$$

where

$$A_r = \begin{bmatrix} -\frac{C_\phi}{I_x} & \frac{m_s g h_s - K_\phi}{I_x} \\ 1 & 0 \end{bmatrix}, \quad B_r = \begin{bmatrix} \frac{1}{I_x} \\ 0 \end{bmatrix}, \quad E_r = \begin{bmatrix} -\frac{m_s h_s}{I_x} \\ 0 \end{bmatrix}.$$

For the rollover mitigation control, LTR index [41, 42] is widely adopted as the evaluation metric for vehicle rollover tendency. Compared with the SSF index discussed in Section 1.2 which considers only geometrical aspects, the LTR is a more proper index to quantify the vehicle

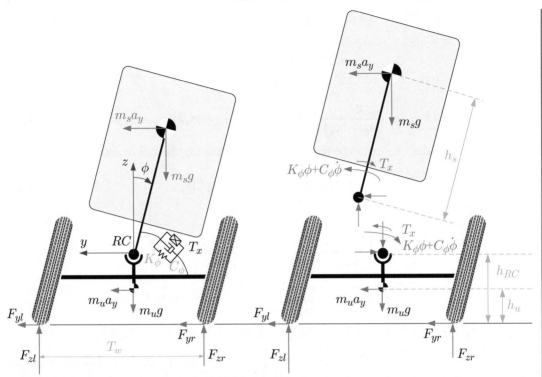

Figure 3.2: Vehicle roll model for NTV.

dynamical stability under active tilting control. By definition [43], LTR can be written by vertical tire forces as

$$LTR \triangleq \frac{F_{zl} - F_{zr}}{F_{zl} + F_{zr}}. \tag{3.10}$$

Intuitively, the vehicle rollover can be defined as a situation when one of the wheels lifts off the ground. In other words, it happens when the vertical load on any of the wheels drops to zero. According to the definitions in Eq. (3.10), LTR index could vary from $[-1, 1]$. For an ideally balanced vehicle with no lateral load transfer, $LTR = 0$.

Since vertical tire forces are not directly measurable on mass-produced vehicles, it is desirable to rewrite the LTR index in terms of measurable parameters and states for control purposes [44, 45]. From the roll model derived in Eq. (3.9), the normal load difference can be written as

$$F_{zl} - F_{zr} = \frac{2}{T_w} \left(I_x \ddot{\phi} + m_u h_u a_y + m_s h'_s a_y - m_s g h_s \phi \right), \tag{3.11}$$

where $h'_s = h_s + h_{rc}$ is the CoG height of the sprung mass.

Substituting the roll dynamics Eq. (3.9) into (3.11), and considering the condition of the static load balance,

$$F_{zl} + F_{zr} = (m_s + m_u) g. \tag{3.12}$$

The LTR index defined in Eq. (3.10) can be re-written as

$$LTR = \frac{2}{T_w (m_s + m_u) g} \left(I_x \ddot{\phi} + m_u h_u a_y + m_s h'_s a_y - m_s g h_s \phi \right). \tag{3.13}$$

Or, equivalently, in the state-space form by substituting the roll dynamics as

$$LTR = C_r X_r + D_r U_r + F_r W_r, \tag{3.14}$$

where

$$C_r = \left[\frac{-2C_\phi}{(m_s + m_u) g T_w} \quad \frac{-2K_\phi}{(m_s + m_u) g T_w} \right]$$
$$D_r = \frac{2}{(m_s + m_u) g T_w}, \quad F_r = \frac{2 (m_s h_{rc} + m_u h_u)}{(m_s + m_u) g T_w}.$$

3.3   INTEGRATED RE-CONFIGURABLE MODEL

Previous sessions developed NTV models in separate lateral and roll planes following the procedure for conventional cars. However, vehicle dynamics in lateral and roll motions are strongly coupled. The planner motion builds up lateral accelerations which serve as disturbances in roll dynamics, as shown in Eq. (3.8), while the roll motion could affect the lateral dynamics by changing tire forces through vertical load distributions as well as camber angles.

An integrated dynamical model is thus proposed to handle such couplings for NTVs. In addition to the re-configurable approach to address different wheel configurations as demonstrated via the generalized double-track model in Section 3.1, incorporation of various actuators such as active steering δ_{cij}, torque vectoring $Q_{cij} > 0$, and differential braking $Q_{cij} < 0$ is proposed using a configuration matrix approach as detailed below.

By combining lateral dynamics in Eq. (3.1) with the roll equations from Eq. (3.9), an integrated re-configurable vehicle model in state-space can be derived as

$$\dot{X} = AX + BRU + EW + G, \tag{3.15}$$

where the full vehicle states are considered as $\left(X = \begin{bmatrix} v & r & \dot{\phi} & \phi \end{bmatrix}^T \right)$; driver's steering maneuver at the front axle is treated as measurable disturbances ($W = \delta_d$). All possibles active steering angles δ_{cij} and torque distributions Q_{cij} on each wheel are considered for a general case as

$$U = \begin{bmatrix} \delta_{cfl} & \delta_{cfc} & \delta_{cfr} & \delta_{crl} & \delta_{crc} & \delta_{crr} \\ Q_{cfl} & Q_{cfc} & Q_{cfr} & Q_{crl} & Q_{crc} & Q_{crr} & T_x \end{bmatrix}^T.$$

To handle various actuators for NTV stability control, a configuration matrix composed of Boolean operators to denote the availability of each actuator [46, 47] is adopted, which can be defined for a general case as

$$R = diag\left(\begin{bmatrix} R_{\delta fl} & R_{\delta fc} & R_{\delta fr} & R_{\delta rl} & R_{\delta rc} & R_{\delta rl} \\ R_{Qfl} & R_{Qfc} & R_{Qfr} & R_{Qrl} & R_{Qrc} & R_{Qrr} & R_{Tx} \end{bmatrix}\right)$$

with $R_{\delta ij}$, R_{Qij} denoting availability of active steering and torque distribution capability at each corner ij, respectively.

The system matrices A, B, E, G shown in Eq. (3.15) can be derived as

$$A = \begin{bmatrix} \dfrac{-\sum\limits_{i,j} \tilde{C}_{aij}}{(M_s+M_u)\cdot u} & \dfrac{-\sum\limits_{i,j} a_i \tilde{C}_{aij}}{(M_s+M_u)\cdot u} - u & 0 & \dfrac{\sum\limits_{i,j} K_{\gamma i} \tilde{C}_{\gamma ij}}{M_s+M_u} \\[1em] \dfrac{-a\sum\limits_{i=f,j} \tilde{C}_{aij}+b\sum\limits_{i=r,j} \tilde{C}_{aij}}{I_z\cdot u} & \dfrac{-\sum\limits_{i,j} a_i^2 \tilde{C}_{aij}}{I_z\cdot u} & 0 & \dfrac{\sum\limits_{i,j} a_i K_{\gamma i} \tilde{C}_{\gamma ij}}{I_z} \\[1em] -\dfrac{M_s H_s \sum\limits_{i,j} \tilde{C}_{aij}}{(M_s+M_u)\cdot u\cdot I_x} & \dfrac{-M_s H_s \sum\limits_{i,j} a_i \tilde{C}_{aij}}{(M_s+M_u)\cdot u\cdot I_x} & -\dfrac{C_\phi}{I_x} & \dfrac{M_s H_s \sum\limits_{i,j} K_{\gamma i} \tilde{C}_{\gamma ij}}{(M_s+M_u)\cdot I_x} + \dfrac{M_s g H_s - K_\phi}{I_x} \\[1em] 0 & 0 & 1 & 0 \end{bmatrix}$$

$$B = \begin{bmatrix} \dfrac{\tilde{C}_{afl}}{M_s+M_u} & \dfrac{\tilde{C}_{afc}}{M_s+M_u} & \dfrac{\tilde{C}_{afr}}{M_s+M_u} & \dfrac{\tilde{C}_{arl}}{M_s+M_u} & \dfrac{\tilde{C}_{arc}}{M_s+M_u} & \dfrac{\tilde{C}_{arr}}{M_s+M_u} \\[1em] \dfrac{a\tilde{C}_{afl}}{I_z} & \dfrac{a\tilde{C}_{afc}}{I_z} & \dfrac{a\tilde{C}_{afr}}{I_z} & \dfrac{-b\tilde{C}_{arl}}{I_z} & \dfrac{-b\tilde{C}_{arc}}{I_z} & \dfrac{-b\tilde{C}_{arr}}{I_z} \\[1em] \dfrac{M_s H_s \tilde{C}_{afl}}{(M_s+M_u)\cdot I_x} & \dfrac{M_s H_s \tilde{C}_{afc}}{(M_s+M_u)\cdot I_x} & \dfrac{M_s H_s \tilde{C}_{afr}}{(M_s+M_u)\cdot I_x} & \dfrac{M_s H_s \tilde{C}_{arl}}{(M_s+M_u)\cdot I_x} & \dfrac{M_s H_s \tilde{C}_{arc}}{(M_s+M_u)\cdot I_x} & \dfrac{M_s H_s \tilde{C}_{arr}}{(M_s+M_u)\cdot I_x} \\[1em] 0 & 0 & 0 & 0 & 0 & 0 \\[1em] 0 & 0 & 0 & 0 & 0 & 0 & 0 \\[1em] 0 & 0 & 0 & 0 & 0 & 0 & 0 \\[1em] -\dfrac{1}{2}\dfrac{T_w}{R_{wfl}I_z} & 0 & \dfrac{1}{2}\dfrac{T_w}{R_{wfr}I_z} & -\dfrac{1}{2}\dfrac{T_w}{R_{wrl}I_z} & 0 & \dfrac{1}{2}\dfrac{T_w}{R_{wrr}I_z} & \dfrac{1}{I_x} \\[1em] 0 & 0 & 0 & 0 & 0 & 0 & 0 \end{bmatrix}$$

$$E = \begin{bmatrix} \dfrac{\sum\limits_{i=f,j} \tilde{C}_{aij}}{M_s+M_u} \\[1em] \dfrac{a\sum\limits_{i=f,j} \tilde{C}_{aij}}{I_z} \\[1em] \dfrac{M_s H_s \sum\limits_{i=f,j} \tilde{C}_{aij}}{(M_s+M_u)\cdot I_x} \\[1em] 0 \end{bmatrix}, \qquad G = \begin{bmatrix} \dfrac{\sum\limits_{i,j} \left(\bar{F}_{yij}-\tilde{C}_{\alpha ij}\bar{\alpha}_{ij}-\tilde{C}_{\gamma ij}\bar{\gamma}_{ij}\right)}{M_s+M_u} \\[1em] \dfrac{\sum\limits_{i,j} a_i \left(\bar{F}_{yij}-\tilde{C}_{\alpha ij}\bar{\alpha}_{ij}-\tilde{C}_{\gamma ij}\bar{\gamma}_{ij}\right)}{I_z} \\[1em] \dfrac{M_s H_s \sum\limits_{i,j} \left(\bar{F}_{yij}-\tilde{C}_{\alpha ij}\bar{\alpha}_{ij}-\tilde{C}_{\gamma ij}\bar{\gamma}_{ij}\right)}{(M_s+M_u)I_z} \\[1em] 0 \end{bmatrix}.$$

3.4 SUSPENSION DESIGNS FOR NTV STABILITY

Among the three major tilting mechanisms reviewed in Chapter 2, the suspension tilting system has a great potential to be adopted for NTVs due to their compact size and re-use of the suspension mechanism which helps to reduce system weight and cost. However, generating pure asynchronous motions of the suspension mechanism along with wheel assemblies are more complicated than visualized in Figure 2.4. Due to manufacturing and maintenance difficulties, a guide-rail structure allowing wheel assemblies to move along its rail (a.k.a. sliding pillar mechanism [48]) is rarely seen on modern automobiles. Instead, various suspension mechanisms have been proposed which introduce lateral and longitudinal wheel motions as the suspension struts are driven for tilting. When suspension motions are relatively large for NTVs during harsh cornering, the impact of suspension kinematics on dynamical behavior is no loner negligible. In this section, a trailing-arm mechanism is adopted as an example for the analysis, and the demonstrated procedure can be implemented on suspension mechanisms for similar NTV applications.

3.4.1 ROLL ANGLE ANALYSIS

This subsection focuses on the kinematics perspectives of additional roll angle changes considering the suspension mechanism during tilting. A tadpole-configured NTV with the trailing arm mechanism is adopted, as shown in Figure 3.3.

Instead of analyzing the vehicle in a tilted position, it would be much easier to visualize and study NTV kinematics with its cabin in an upright pose and a "rotating" ground plane. The plane can be constructed, for any cabin pose, using contact points at three wheels. By comparing the angle between normal vectors of the new contact plane with the original (un-tilted) one, the vehicle rotation as a consequence of suspension movements can then be analyzed [24].

The original ground plane $OXYZ$ is adopted as the global coordinate system with OZ pointing upward. The un-tilted pose of the NTV can be defined with OXZ be the symmetrical plane of the vehicle; OX pointing forward; and OZ across the CoG of the vehicle. The contact points are assumed to be located at the bottom of wheels. Points A, B, C denote the nominal contact points at front-left, front-right, and rear wheels for the un-tilted pose, while A', B', C' denotes their new positions after enforcing suspension movements for tilting. The normal directions of the original and new ground planes are denoted as $\hat{n}_1 \equiv \begin{bmatrix} 0 & 0 & 1 \end{bmatrix}^T$ and $\hat{n}_2$ accordingly.

Given the nominal track width at front axle T_{w0}, and axle to CG distances (a, b), position vectors for the original contact points A, B, C are

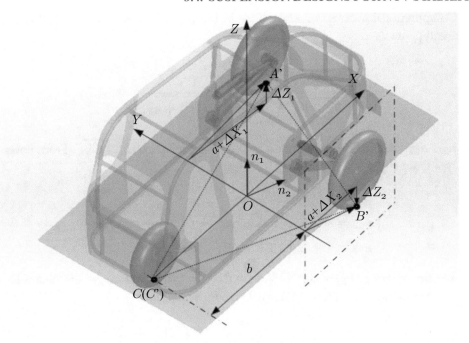

Figure 3.3: Kinematic analysis of NTV with trailing arm mechanisms [24]. (Image used with permission.)

$$\overrightarrow{OA} = \left[\begin{array}{ccc} a & +\dfrac{T_{w0}}{2} & 0 \end{array}\right]^{T}$$
$$\overrightarrow{OB} = \left[\begin{array}{ccc} a & -\dfrac{T_{w0}}{2} & 0 \end{array}\right]^{T} \qquad (3.16)$$
$$\overrightarrow{OC} = \left[\begin{array}{ccc} -b & 0 & 0 \end{array}\right]^{T}.$$

For a general tilting scenario, Δz_1 and Δz_2 are applied, respectively, as the vertical displacements of the two front wheels. Since the wheels are constrained by the trailing arm mechanism, no lateral displacement is produced, but both wheels move Δx_1 and Δx_2, respectively, in the longitudinal direction. The global position vectors of the new contact points A', B', C' are written as

$$\overrightarrow{OA'} = \overrightarrow{OA} + \left[\begin{array}{ccc} \Delta x_1 & 0 & \Delta z_1 \end{array}\right]^{T}$$
$$\overrightarrow{OB'} = \overrightarrow{OB} + \left[\begin{array}{ccc} \Delta x_2 & 0 & \Delta z_2 \end{array}\right]^{T} \qquad (3.17)$$
$$\overrightarrow{OC'} = \overrightarrow{OC}.$$

Computational geometry approach could be adopted to solve the normal vector of the new ground plane $\hat{n}_2$ as

$$\hat{n}_2 = \frac{\overrightarrow{C'B'} \times \overrightarrow{C'A'}}{|\overrightarrow{C'B'}||\overrightarrow{C'A'}|} = \frac{1}{\sqrt{n_{21}^2 + n_{22}^2 + n_{23}^2}} \begin{bmatrix} n_{21} \\ n_{22} \\ n_{23} \end{bmatrix}, \tag{3.18}$$

where n_{21}, n_{22}, n_{23} are the components of the cross product $\overrightarrow{C'B'} \times \overrightarrow{C'A'}$. From position vectors solved from Eqs. (3.16) and (3.17), this leads to

$$n_{21} = -\frac{T_{w0}}{2}(\Delta z_1 + \Delta z_2)$$

$$n_{22} = -(a + b + \Delta x_2)\Delta z_1 + (a + b + \Delta x_1)\Delta z_2$$

$$n_{23} = \frac{T_{w0}}{2}(2a + 2b + \Delta x_1 + \Delta x_2).$$

Rotational motion (ϕ) of the NTV, with the proposed "rotation ground" formulation, can be determined by the angle between normal vectors $\hat{n}_1$ and $\hat{n}_2$ as

$$\phi = \arccos(\hat{n}_1 \cdot \hat{n}_2) = \arctan\left(\sqrt{\frac{n_{21}^2 + n_{22}^2}{n_{23}^2}}\right). \tag{3.19}$$

The instantaneous rotation axis $\hat{e}_r$ is found by the cross product of normal vectors as

$$\hat{e}_r = \hat{n}_2 \times \hat{n}_1 = \frac{1}{\sqrt{n_{21}^2 + n_{22}^2 + n_{23}^2}} \begin{bmatrix} n_{22} \\ -n_{21} \\ 0 \end{bmatrix}. \tag{3.20}$$

It is shown in Eq. (3.20) that the rotation axis has zero component in the global coordinate system, which indicates no yaw motion can be generated by arbitrary movements of trailing arm suspensions [24]. The projected rotation along OX and OY directions can be interpreted as the tilting and pitch motion, respectively. To eliminate the undesired pitch, the term n_{21} in Eq. (3.20) has to be zero, which leads to

$$\Delta z_1 + \Delta z_2 = 0. \tag{3.21}$$

It is suggested in Eq. (3.21) that, for trailing arm mechanisms to generate a pure tilting motion without introducing pitch, asynchronous motions (same magnitudes but reverse directions) of wheel modules in the vertical direction are desired. This agrees with the intuition that any synchronous wheel movement will lift the front axle and end up with pitch motion of the full vehicle.

Imposing this condition in the rotation angle solved from Eq. (3.19) leads to a simplified pure tilting angle ϕ_x as

$$\phi_x = \phi|_{\Delta z_1 = +\Delta z, \, \Delta z_2 = -\Delta z} = \arctan\left(\frac{2\Delta z}{T_{w0}}\right). \tag{3.22}$$

Under a pure-tilting condition, as suggested by Eq. (3.22), longitudinal movements (Δx_1 and Δx_2) introduced by the trailing arm suspensions will not affect the tilting angle. The angle can be determined by using the simplified roll plane model as adopted in previous sections. However, once the no-pitch condition Eq. (3.21) does not hold, the general rotation angle ϕ solved from Eq. (3.19) should be considered, and both longitudinal and vertical movement of the wheels can affect the final tilting angle.

3.4.2 LOAD DISTRIBUTION ANALYSIS

Apart from direct impacts on motion states (e.g., ϕ), suspension kinematics could also change normal force distributions of NTVs due to movements of contact patches as imposed by suspension tilting mechanisms. LTR index derived in Section 3.2 is adopted as the stability measure to analyze the impact of wheel movements on vehicle rollover stability. A tadpole-configured NTV with trailing-arm suspension is illustrated in Figure 3.4 for this analysis.

A quasi-static assumption is adopted which applies inertia forces ma_x, ma_y at center-of-gravity of the NTV, and the tilting angle is set to its steady-state value ϕ_x. To balance positive a_y disturbances, the vehicle should lean to the left, with its left wheel being lifted and right wheel pushed downward, as shown in the figure.

Using D'Alembert's principle, the equilibrium considering wheel movements imposed by suspension mechanisms is derived as

$$\begin{cases} F_{z1} + F_{z2} + F_{z3} = mg \\ mgh_0 \sin(\phi_x) + ma_y h_0 \cos(\phi_x) + (F_{z1} - F_{z2})\frac{T_w}{2} = 0 \\ ma_x h_0 \cos(\phi_x) + F_{z1}(a + \Delta x_1) + F_{z2}(a + \Delta x_2) - F_{z3}(b) = 0. \end{cases} \tag{3.23}$$

Normal loads on each tire can then be solved from Eq. (3.23) as

$$\begin{cases} F_{z1} = \frac{b}{2l}mg - \frac{mh}{2l}a_x - \frac{l_0 + \Delta x_2}{l}\frac{mh}{T_{w0}}a_{ynb} \\ F_{z2} = \frac{b}{2l}mg - \frac{mh}{2l}a_x + \frac{l_0 + \Delta x_1}{l}\frac{mh}{T_{w0}}a_{ynb} \\ F_{z3} = \frac{a + (\Delta x_1 + \Delta x_2)/2}{l}mg + \frac{mh}{l}a_x - \frac{\Delta x_1 - \Delta x_2}{l}\frac{mh}{T_{w0}}a_{ynb} \end{cases}, \tag{3.24}$$

where l_0 is the nominal wheelbase $l_0 = a + b$; l is the actual wheelbase considering wheel longitudinal movements during tilting; and a_{ynb} is the unbalanced lateral acceleration by subtracting the gravity-balanced terms:

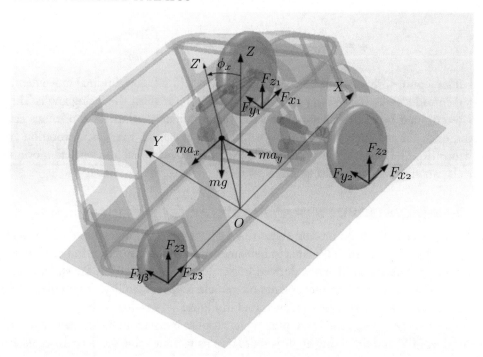

Figure 3.4: Force analysis of NTV in tilting [24]. (Image used with permission.)

$$l = l_0 + \frac{\Delta x_1 + \Delta x_2}{2}$$
$$a_{ynb} = a_y \cos(\phi_x) + g \sin(\phi_x).$$

(3.25)

It is shown in Eq. (3.24) that wheel motions enforced by suspension mechanisms do effect normal load distributions. To quantify its impact, LTR [49, 50] can be solved using normal forces determined on front wheels as

$$LTR = \frac{(\Delta x_1 + \Delta x_2)/2 + l_0}{-(\Delta x_1 - \Delta x_2)/2 + p}$$

(3.26)

with

$$p = \frac{1}{2} \frac{\frac{a_x}{g} h - b}{\frac{a_{ynb}}{g} \frac{h}{T_{w0}}}.$$

Compared with this general LTR index considering the impacts of suspension kinematics, the classical roll-plane LTR index in previous sections can be treated as a special case with $\Delta x_1 = \Delta x_2 = 0$.

3.4.3   APPLICATION IN SUSPENSION MECHANISM DESIGN

Based on the kinematic (Section 3.4.1) and dynamic (Section 3.4.2) analysis of NTV suspensions, optimizations of the mechanism by considering its impact on NTV behavior could be applied to enhance vehicle roll stability even further and reduce the tilting effort.

For a design example with trailing-arm suspension, since magnitudes of wheel longitudinal movements (Δx_1, Δx_2) should be less than the nominal wheelbase l_0, signs of the nominator and denominator terms of the LTR index in Eq. (3.26) are determined as [24]:

$$\begin{cases} sign\left((\Delta x_1 + \Delta x_2)/2 + l_0\right) = sign\left(l_0\right) > 0 \\ sign\left(-(\Delta x_1 - \Delta x_2)/2 + p\right) = sign\left(p\right) = -sign(a_y). \end{cases} \tag{3.27}$$

By observing the general rollover index from Eq. (3.26), a necessary condition for Δx_1 and Δx_2 to improve the vehicle roll stability by minimizing the magnitude of LTR index is

$$\begin{cases} \Delta x_1 + \Delta x_2 \leq 0 \\ sign(a_y)\left(\Delta x_1 - \Delta x_2\right) \geq 0. \end{cases} \tag{3.28}$$

Since lateral acceleration a_y points toward the inner bend of the curve, the two front wheels can be denoted as inner and outer wheels, respectively. According to vehicle cornering directions, the necessary condition in Eq. (3.28) can be simplified as

$$\begin{cases} \Delta x_{in} + \Delta x_{out} \leq 0 \\ \Delta x_{in} - \Delta x_{out} \geq 0 \end{cases} \Leftrightarrow \begin{cases} \Delta x_{out} \leq 0 \\ |\Delta x_{in}| \leq -\Delta x_{out}, \end{cases} \tag{3.29}$$

where Δx_{in} and Δx_{out} denote the wheel longitudinal movement on inner- and outer-side, respectively.

This necessary condition is also visualized in Figure 3.5. In order to realize the preferred wheel longitudinal movement and improve the rollover stability during tilting, the outer-wheel is expected to move backward, while the inner-wheel is free to move in both forward and backward directions, with the magnitude bounded by that of the outer wheel. Such conditions could then be formulated in a suspension design optimization process as detailed in [24].

3.5   CONCLUSIONS

This chapter studies the vehicle dynamics of tilting vehicles. Starting with vehicle modeling in lateral and roll planes, a re-configurable structure which handles various NTV configurations is proposed, and an integrated NTV model is suggested. LTR as a roll stability measure is derived for control implementations in the next chapter. The impact of suspension mechanism on NTV dynamics is discussed in detail, which suggests a proper mechanism design could be utilized to enhance NTV stability even further.

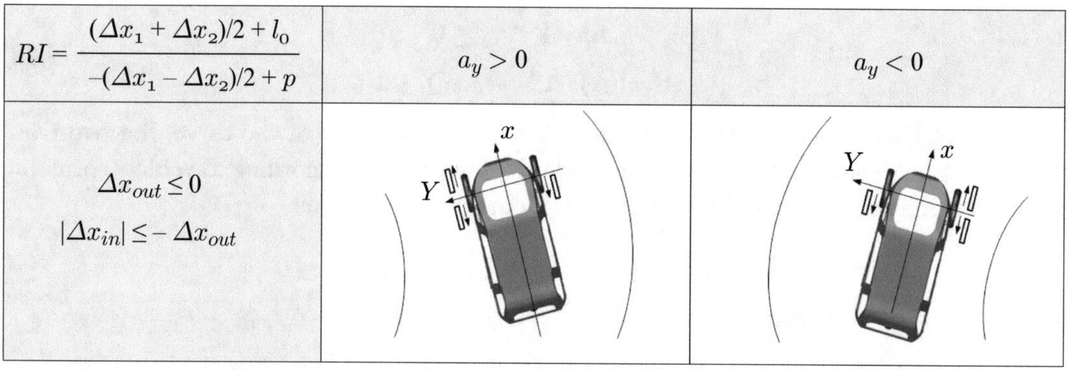
$RI = \frac{(\Delta x_1 + \Delta x_2)/2 + l_0}{-(\Delta x_1 - \Delta x_2)/2 + p}$
$a_y > 0$
$a_y < 0$
$\Delta x_{out} \leq 0$
$|\Delta x_{in}| \leq -\Delta x_{out}$
x
Y
Y
x

Figure 3.5: Preferred wheel longitudinal motion for vehicle roll stability improvement [24]. (Image used with permission.)

CHAPTER 4

Tilting Vehicle Control

This chapter presents the controller design for NTVs. Based on the vehicle model previously developed, and the adoption of LTR as a measure of vehicle rollover tendency, a safe roll envelope is suggested as the activation criteria for active tilting control. Instead of tracking zero LTR index which could be too conservative, a more energy-efficient approach to activate the tilting actuator only when the vehicle states are leaving the safe boundary is suggested.

The motivation and derivation of the safe roll envelope is presented after a brief literature review of vehicle roll control strategies. Various representations including the feed-forward tilting moment, the steady-state roll angle, and roll state constraints are derived as specific forms of the general LTR expression for control purposes.

Both feed-forward and feedback controller designs are presented to demonstrate the usage of simplified roll envelopes in controller developments. The feed-forward approach is featured by its easy implementation with no prerequisites on direct roll state measurements, while the feedback one, by utilizing the extra information, performs better under harsh maneuvers. The feedback controller also demonstrates an integrated controller design by combining the lateral and roll dynamics to form a holistic control architecture. As a direct benefit, better distributions of control efforts for safety and efficiency can be achieved. The overall system robustness against failure also gets enhanced.

To utilize the roll envelope in its full state-constraints form, a model predictive control approach is adopted. The significance of the constraint formulation to vehicle transient performance enhancement is discussed via the analysis of the non-minimum phase issue in active tilting control. The model predictive control (MPC) approach for rollover mitigation provides a good balance between quick system responses and small control overshoots. Further integrations of the proposed roll envelope with the lateral stability envelope leads to an integrated envelope approach for vehicle stability control. Optimum control distributions, the flexibility in actuator selections, as well as enhancements of system robustness are demonstrated via the re-configurable holistic MPC approach.

4.1 CHALLENGES IN TILTING CONTROLLER DESIGN

For the control of NTVs, major challenges such as the following are considered.

Energy Efficiency in Control Strategy

As reviewed in one of the challenges for tilting mechanism designs, efficient tilting control is believed to be important for the success of NTVs. Tilting control as an active safety system should not consume too much energy, otherwise it conflicts with the original idea to build compact and energy-efficient cars for urban use. Active suspensions systems [41], as an example from conventional vehicles, demonstrate great improvements on road holding as well as riding comport, but are only seen on limited luxury vehicles due to their high fuel consumption. Compared with the conventional tilting control algorithms which try to perfectly balance the vehicle like a motorcycle, the envelope-based roll stability control is proposed in this chapter. By realizing that vehicle track-width passively helps to resist some portion of unbalanced lateral accelerations without causing rollover problems, an envelope control scheme based on LTR thresholds is proposed to tilt the vehicle in an energy-efficient manner.

Optimum Distribution of Control Efforts

Apart from the adoption of envelope approaches for an energy-efficient tilting control, an integrated vehicle controller provides an alternative solution by systematically optimize the control effort distributions among all available actuators. On conventional cars, previous researches for integrated vehicle controls either ignore the roll dynamics [51, 52, 52] or separately control the roll from the lateral and longitudinal motions [53, 54]. A promising direction shown in NTV literature is to control vehicle roll dynamics via active steering [18–20] which are known as STC/SDTC methods. However, these endeavours only consider the coordination of steering and tilting actuators, which could be generalized to form a holistic vehicle control structure to fully utilize other actuators like torque vectoring and differential braking. Apart from seeking a minimum-effort action, the integrated control scheme also promotes systematic considerations of system constraints, which greatly enhance the NTV safety as well.

Non-Minimum Phase in Active Roll Motion Control

Applying the tilting moment can deteriorate the LTR index in transient, which could be troublesome for NTV control if the use of such actuators are not properly designed. To handle this non-minimum phase issue without being too conservative in control actuation, the model predictive controller with the roll envelope constraint is suggested. The predictive feature of the control architecture can be used to detect possible envelope violations at an early stage to trigger the rollover mitigation control proactively, and quantify the LTR index overshoot for better control decisions if it is unavoidable.

Flexible Architecture

Another challenge in controller design for NTV is the various wheel and actuator configurations, as shown in Table 1.1 and discussed in Section 3.1. A properly designed controller for one specific tilting vehicle might not work for others due to configuration changes. To address this, re-configurable control architectures are demonstrated with the proposed sliding mode control (SMC) and model predictive control (MPC) designs. The suggested framework not only improves flexibility of the control system but also makes it more robust against actuator failure since the control effort can be optimally redistributed online to react to detected failures.

4.2   REVIEW OF ACTIVE ROLL CONTROL APPROACHES

This section reviews the technique for active roll motion controls. Since the conventional road vehicles with high CoG share the same rollover problem, the review starts with the active roll control for SUVs and trucks, and then focused on those specific controller designs for narrow tilting vehicles.

4.2.1   ACTIVE ROLL CONTROL FOR CONVENTIONAL VEHICLES

The vehicle rollover is known as a major cause of severe and fatal injuries in traffic accidents. According to the research by the National Highway Traffic Safety Administration (NHTSA) [55], rollover accidents are ranked as the second most dangerous form of accident in the United States, after head-on collisions. Statistics also suggest that although a small portion of all accidents involves rollover, they constitute a disproportionately large portion of fatal ones [56].

Large vans and sport utility vehicles are prone to rollover accidents due to their high CoG positions [51, 57], which could cause rollover problems before losing grip [2]. For rollover mitigation controls, most of the existing researches are focused on un-tripped rollovers [52, 58], which are caused by extreme maneuvers like excessive speed cornering or harsh obstacle avoidance maneuvers. Practices to improve vehicle rollover stability can be categorized as:

- direct control of vehicle roll motions such as active suspensions, active anti-roll bars, and active stabilizers; and

- indirect roll control via regulations of vehicle lateral accelerations such as differential braking and active steering.

Among those approaches that utilize active suspensions for rollover mitigation, LTR is adopted as an evaluation criteria for control performances [56, 57]. Reported drawbacks of the direct control approaches are that it influences the lateral stability of the vehicle and could causes an over-steer characteristic [59].

The most common approach for indirect control of rollover is based on the reduction of the lateral acceleration by decreasing the yaw rate. This approach has been implemented through differential braking and active steering [60, 61]. The limitation of the approach is a possible loss of maneuverability, which may cause another accident [59, 62]. Some studies have been conducted to solve this problem for having both rollover prevention and good lateral stability [39, 63].

4.2.2   ACTIVE ROLL CONTROL FOR NARROW TILTING VEHICLES

For NTVs, which are more like motorcycles than conventional cars, an active rollover mitigation is even more crucial. Compared with two-wheelers, the increased system weight along with the enclosed cabin of NTVs makes the balancing control much more difficult to be executed by human drivers. Such vehicles are also expected to be operated by drivers without sufficient skill and experience to lean the vehicle when cornering. Consequently, active tilting control systems have been developed to keep vehicles in balance without the driver's direct intervention. There are mainly three different types of tilting control schemes for NTVs known as Direct Tilt Control (DTC), Steering Tilt Control (STC), and a combination of Steering and Direct Tilt Control (SDTC).

Direct Tilting Control (DTC)

For DTC systems, tilting actuators rotate the body relative to the chassis, and the steering system is used exclusively for lateral control [7]. The desired tilting angle is first determined by considering the vehicle speed, the road curvature and vehicle running states. The controller then tracks the desired angle using tilting actuators. Strategies to determine an appropriate tilting angle are crucial to DTC system designs.

The common method to calculate the desired tilting angle is based on the idea of balancing the centrifugal force during steady-state cornering [7] like motorcycles. The idea is adopted in [2, 60, 61] for DTC system designs. A more complicated version is suggested in [7], which takes the vehicle lateral acceleration and gyroscopic moments of rotating wheels into considerations. Authors claim that incorporation of the gyroscopic term ensures no tilting torques are required from actuators in steady state.

Steering Tilting Control (STC)

For STC systems, there are no exclusive actuators for tiling motion controls. Instead, the steering effort is used to track the given trajectory as well as balancing the vehicle.

Based on steady-state steering conditions and linear tire model assumptions, the steering angle can be written as a function of the desired tilting angle, vehicle running speed, and road curvature. However, determination of a desired steering angle which stabilizes the vehicle while at the same time tracks the trajectory is no easy task. Opposite steering efforts are required for curve negotiation and tilting [7].

To resolve this, driver's steering intention is no longer fed to the steering system directly. Instead, it will first be used to determine a desired tilting angle, which is then tracked using steering control efforts. As the steady-state yaw rate and the tilting angle are coupled, by tracking a certain tilting angle derived from driver's intention, the suggested steering control inherently tracks a reference yaw rate for trajectory following.

Integrated Steering and Direct Tilting Control (SDTC)

DTC system is more stable at lower speeds compared to the STC system, while STC is more efficient at higher speed and requires less torque in transient [64–66]. To incorporate the benefit of both, a hybrid approach is proposed in which DTC is more dominant at lower speeds and then STC takes over at higher speeds.

A switching approach based on vehicle speed is suggested by So and Karnopp [67]. Snell [68] develops a combined SDTC control system where both actuators are active all the time. Kidane et al. also proposes an SDTC system with both DTC and STC operating and a sinusoidal weighting function is used to ensure a smooth switch [7, 69].

4.3   ENVELOPE-BASED ROLL STABILITY CONTROL

This section introduces the idea of roll envelope control which is adopted throughout the chapter. The motivation of the envelope-based control approach is discussed, first followed by several forms of implementations of the roll envelope with different complexities.

4.3.1   MOTIVATION FOR ENVELOPE-BASED CONTROL

For rollover mitigation studies, one of the most important steps is to adopt a good measure for the roll stability. When vehicles are approaching their tipping limits, one of the tires should lose contact with the ground, and the vertical tire force drops to zero. This process could be well captured by the LTR index derived in Chapter 3. For a vehicle with well-balanced loads, the LTR index is close to zero, while under harsh rollover scenarios, the LTR index approaches ± 1 [47]. To avoid dangerous situations, it is thus natural to maintain a small LTR index as the control objective for rollover mitigation.

Previous studies for tilting vehicle control [7, 21, 41] attempt to regulate LTR by treating the tilting vehicle as a motorcycle with zero track width, and resultant tilting controllers try to balance the centrifugal force through frequent active tilting during the operation. However, this seems to be unnecessary, as the existence of the vehicle track width can help to endure some unbalanced lateral accelerations, as illustrated in Figure 4.1. The tilting strategy which tries to eliminate the LTR and fully balances the vehicle might be too conservative. Frequent activation of the tilting control system could consume too much energy and conflicts with the original idea to build energy-efficient compact vehicles [70].

By accepting the LTR index does not need to be kept at zero but should only be bounded within a safe envelope defined by the threshold LTR_{lim}, the tilting actuator can be inactive when the vehicle is driven with moderate maneuvers at lower speeds. For harsh maneuvers, the tilting effort can also be reduced since the target is not to perfectly balance the vehicle but only to bring the system back to the safe envelope. Much control effort can be saved when the vehicle is within the safe threshold (LTR_{lim}) as demonstrated in [47], which improves the energy efficiency of the vehicle.

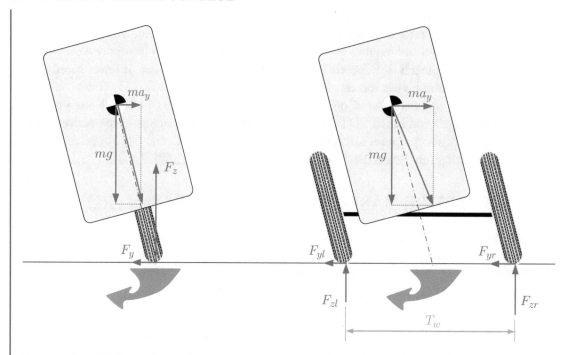

Figure 4.1: Vehicle model with zero and non-zero track width.

Based on the LTR derivation in Eq. (3.13) from the roll dynamics model, the full LTR expression is shown to be composed of transient ($\dot{\phi}$) as well as steady-state (ϕ) terms. Considering NTVs are usually operated at lower speed with less harsh maneuvers, ignoring their transient terms could be a fair approximation, and the resultant controller structure could be much simplified. For the DTC and STC schemes reviewed in Section 4.2, the desired tilting angles are derived under such assumptions. However, keeping both transient and steady-state terms leads to the implementation of the roll envelope as state constraints, which requires more advanced control structure and more computational power to handle. Both forms are derived and discussed in detail below.

4.3.2   STEADY-STATE ROLL MOMENT ENVELOPE

Substituting roll dynamics Eq. (3.8) into the LTR expression derived in Eq. (3.13), and considering steady-state conditions with small angle assumptions leads to

$$LTR_{ss} = k_a a_y + k_t T_x, \tag{4.1}$$

where

$$k_a = \frac{2mh_0}{T_w mg} + \frac{2m_s h_s}{T_w m}\frac{m_s h_s}{K_\phi - m_s g h_s};$$

$$k_t = \frac{-2m_s h_s}{T_w m}\frac{1}{K_\phi - m_s g h_s}.$$

The LTR index in the form of Eq. (4.1) could be used to determine whether to apply the control moment according to a_y measurement and the LTR_{lim} threshold. For a proactive control action in feed-forward designs, lateral acceleration a_y could be estimated based on driver's steering wheel input δ leads to

$$LTR_{ss} = k_a \frac{u^2}{l_0 + K_{us} u^2}\delta + k_t T_x, \tag{4.2}$$

where K_{us} is the vehicle understeer gradient coefficient.

For a given steady-state roll envelope specified by the threshold LTR_{lim}, Eq. (4.2) can be used to calculate the steady-state torque request (T_x), based on driver's steering maneuver (δ) and vehicle operational speed (u) as

$$T_x = \begin{cases} 0 & k_a \frac{u^2}{l_0 + K_{us} u^2}\delta \leq LTR_{lim} \\ \frac{1}{k_t}\left(LTR_{lim} - k_a \frac{u^2}{l_0 + K_{us} u^2}\delta\right) & k_a \frac{u^2}{l_0 + K_{us} u^2}\delta > LTR_{lim}. \end{cases} \tag{4.3}$$

The piece-wise function in Eq. (4.3) demonstrates the idea of proposed envelope control: by comparing the estimated roll stability measure with a pre-designed threshold, the activation condition of roll mitigation control can be tuned by proper envelope selections in control schemes.

4.3.3   STEADY-STATE ROLL ANGLE ENVELOPE

Previous section derives a steady-state roll envelope from a torque perspective for feed-forward tilting controllers. However, information collected from motion sensors on tilting vehicles can greatly enhance the control performance. For this reason, the steady-state roll envelope is also derived from a motion perspective in this section. The resultant steady-state roll angle based on the envelope approach can be used for a more precise control of vehicle roll motions.

Under steady-state conditions, the transient terms in the LTR expression are ignored. Denoting the threshold for LTR index as LTR_{lim}, the desired tilting angle derived from the roll envelope can be written as

$$\phi_{des} = \begin{cases} \frac{mT_w}{2m_s h_s}sign\left(a_y\right)LTR_{lim} - \frac{mh_0}{m_s h_s g}a_y & |a_y| \geq a_y^* \\ \phi_{passive} & \text{otherwise,} \end{cases} \tag{4.4}$$

where the lateral acceleration threshold a_y^* to activate the tilting control is solved as

$$a_y^* = \frac{mT_w}{2m_s h_s\left(\frac{m_s h_s}{K_\phi - m_s g h_s} + \frac{mh_0}{m_s g h_s}\right)}LTR_{lim}. \tag{4.5}$$

It should be mentioned that by setting a zero track-width (i.e., $T_w = 0$), or a zero rollover index (i.e., $LTR_{lim} = 0$), the desired tilting angle derived from the envelope approach can be reduced to its the special form adopted by conventional tilting controls ($\phi_{des} \approx -a_y/g$). The general expression given in Eq. (4.4) also considers the sprung and un-sprung mass distribution, which could be very different for full tilting and partial tilting vehicles, as shown in Table 1.1. It is also revealed by Eq. (4.5) that smaller track-width (T_w), as well as a lower rollover index threshold (LTR_{lim}), could activate the tilting control at an earlier stage [70].

A major benefit of the proposed envelope approach for tilting control is the energy saving. To demonstrate this, the energy consumption on tilting actuators (J_{tilt}) to achieve the desired roll stability can be calculated using the inverse roll dynamics from Eq. (3.9), as detailed in [70]. Figure 4.2 illustrates the desired tilting angle and energy consumption under a quasi-static process for various LTR index thresholds. It can be seen that the energy consumption increases quadratically as the increase of the lateral acceleration. The energy saving with a higher LTR threshold is majorly due to the "delayed" activation, which applies control efforts only under more severe disturbances. The figure also shows that, maintaining a zero rollover index could be challenging for large a_y disturbances. The suspension mechanism, as well as the actuator, needs to be properly designed to allow the excessive roll motion [70].

The energy consumption as a result of the suggested roll envelope approach is very dependent on the driving style. Adopting the steady-state lateral acceleration in the feed-forward form, as shown in Eq. (4.2), the desired tilting angle in Eq. (4.4) can be represented as a map-

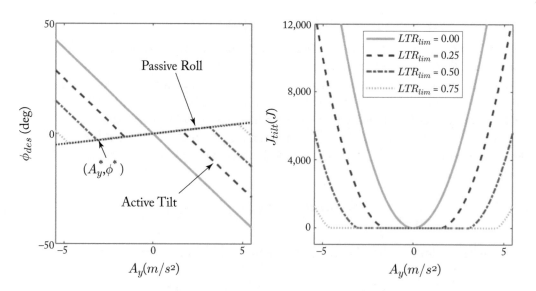

Figure 4.2: Desired tilting angle and energy consumption.

ping dependent solely on vehicle longitudinal speed, steering angle, and threshold of the LTR index. The contour of the desired roll angle with different LTR_{lim} is visualized in Figure 4.3.

The lighter color in the figure denotes smaller tilting angles, while the white area represents the scenarios where the rollover index is below the threshold, and no active tilting control is required. As the increase of the LTR threshold, the area represents scenarios that require no

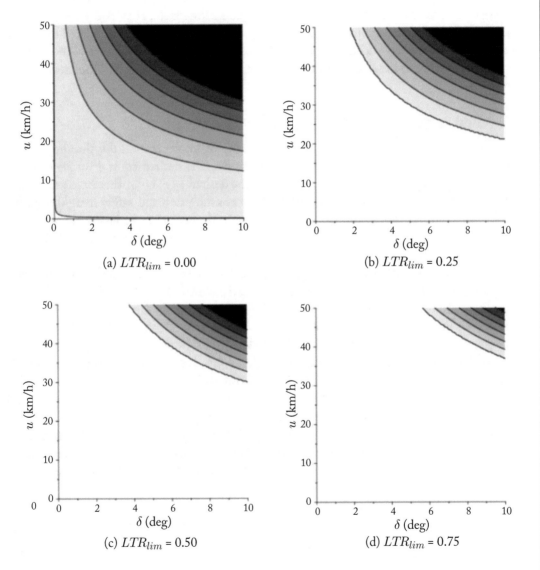

(a) $LTR_{lim} = 0.00$

(b) $LTR_{lim} = 0.25$

(c) $LTR_{lim} = 0.50$

(d) $LTR_{lim} = 0.75$

Figure 4.3: Contour plot of desired tilting angles with various LTR thresholds (lighter color denotes smaller angles).

tilting control grows quickly, while the dark area, which denotes the situations with maximum tilting angles, shrinks to the upper right corner. For a conservative case when the desired LTR index is zero (Figure 4.3a), tilting actuators are activated all the time. The envelope-based tilting control strategy (Figure 4.3b–d) is only expected to deal with rollover tendency at higher speeds with harsher maneuvers.

Although a higher threshold value helps to improve the energy efficiency, determination of a proper threshold LTR_{lim} also needs to consider conflicting factors like safety margin, riding comfort, and driving sensations. All tuning for the threshold need to be validated by driving simulators [71] or tests to achieve a balanced solution. Drivers might also be allowed to adjust such settings from the dashboard to meet their specific driving style and the fuel economy preferences.

4.3.4 ROLL ENVELOPE AS STATE CONSTRAINTS

The previous two sections demonstrate the possibility of implementing the threshold of roll envelope as steady-state torque inputs or motion states. The desired control torque or tilting angle in the form of Eqs. (4.3) and (4.4) considers the desired LTR index threshold as a tunable parameter, which compromises between the energy consumption and safety margin improvements [47]. However, since both expressions are derived under the steady-state assumptions, the transient behavior is not captured. To improve roll control performances, especially under harsh scenarios, a full consideration of both steady-state and transient performances is desired.

With the LTR index written in the form of Eq. (3.14), the prediction of rollover tendency is possible with available roll states measurements and estimations [72]. Furthermore, with the given tilting moment inputs (U_r) and measured disturbances (W_r), the LTR index can be represented as a linear combination of roll states (X_r). If plotted in the roll phase plane ($\phi - \dot{\phi}$), the contour of LTR indices is a set of sloped lines determined by the suspension properties. Two sets of LTR-index boundaries are illustrated in Figure 4.4. The sloped dash-dot lines stand for the desired roll envelope (e.g., $LTR_{\text{lim}} = 0.5$), while the dash lines represent the hard boundary for the rollover situation ($LTR_{\text{lim}} = 1.0$). Horizontal lines in the phase plot represents the roll rate limits $\dot{\phi}_{\text{lim}}$ derived from actuator restrictions as well as passenger comfort requirements.

Compared with roll control approaches which track either zero roll angle (y-axis in the phase plane) or zero rollover index (the line with $LTR_{\text{lim}} = 0.0$), the envelope approach saves the control effort by only maintaining the roll states within the envelope area to achieve the required level of stability.

The equilibrium point for the roll dynamics can be solved from Eq. (3.9) as

$$\phi_{ss} = \frac{-m_s h_s a_y + T_x}{K_\phi - m_s g h_s}. \tag{4.6}$$

When there are no lateral disturbances nor control torques, the passive system should converge to the origin with various initial states located on the rollover index boundaries, as

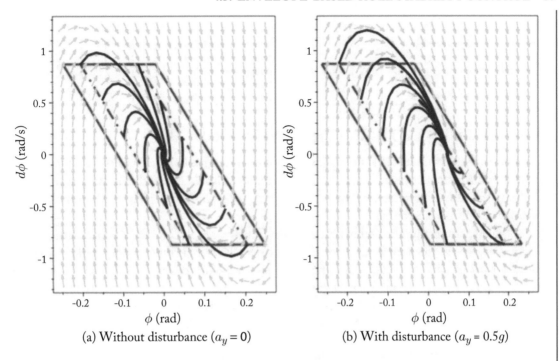

(a) Without disturbance ($a_y = 0$)   (b) With disturbance ($a_y = 0.5g$)

Figure 4.4: Open-loop vehicle phase portrait in roll plane.

shown in Figure 4.4a. However, as the lateral acceleration increases, the equilibrium point will move along the horizontal axis which makes the vehicle less stable, as shown in Figure 4.4b. Some trajectories cross the boundary of rollover index thresholds, which should be mitigated by proper controls.

It should also be noted that the static equilibrium point shown in Eq. (4.6) might not be a good measure to determine whether to activate the roll control. As can be seen in Figure 4.4b, although the static equilibrium point for this case stays inside the desired envelope, there might be transient dynamics causing the states to leave the safe envelope. A model-based approach is adopted to predict the development of LTR index during run-time and activates the envelope control accordingly. The safe roll envelope defined and visualized above can be written as

$$M_{ro}X_r^{(k)} + N_{ro}U_r^{(k)} \leq L_{ro}, \tag{4.7}$$

where

$$M_{ro} = \begin{bmatrix} +C_r \\ -C_r \\ \begin{bmatrix} +1 & 0 \end{bmatrix} \\ \begin{bmatrix} -1 & 0 \end{bmatrix} \end{bmatrix}, N_{ro} = \begin{bmatrix} +D_r \\ -D_r \\ O_{2\times1} \\ O_{2\times1} \end{bmatrix} L_{ro} = \begin{bmatrix} LTR_{\lim} \\ LTR_{\lim} \\ \dot{\phi}_{\lim} \\ \dot{\phi}_{\lim} \end{bmatrix} - \begin{bmatrix} +F_r \\ -F_r \\ 0 \\ 0 \end{bmatrix} W_r^{(k)}.$$

4.4   SIMULATION ENVIRONMENT SETUP

To evaluate the performance of the suggested controller designs in the rest of this chapter, especially with the introduction of the active tilting actuators, simulation studies are implemented in CarSim. The software package is a well-known simulation platform widely-used by automotive industry as well as researchers.

Since there are no validated NTV models in CarSim, a high CoG SUV model with 15 mechanical degrees of freedom is adopted. The vehicle track width is intentionally reduced to represent a narrow urban vehicle. A tire model based on lookup tables is used, which considers the longitudinal force, lateral force, aligning moments, and overturning moments as functions of slip, load, and camber. The vehicle parameters for the simulation are listed in Table 4.1 and various actuator configurations are considered in different simulated scenarios.

Table 4.1: Parameters of the CarSim vehicle model

Symbol	Parameter Description	Value	Unit
m_s	Sprung mass	1590	kg
m	Vehicle mass	1830	kg
I_x	Roll inertia	894.4	$kg \cdot m^2$
I_z	Yaw inertia	2687.1	$kg \cdot m^2$
a	Front axle to CoG distance	1.18	m
b	Rear axle to CoG distance	1.77	m
h_s	CoG to roll center distance	0.72	m
h_u	Un-sprung mass CoG height	0.2	m
T_w	Vehicle track width	1.2	m
K_ϕ	Effective roll stiffness	81363	$N \cdot m/rad$
C_ϕ	Effective roll damping	4432	$N \cdot m/rad$

Three different controller designs will be presented based on the different forms of the roll envelope derived in Section 4.3.

A feed-forward tilting control is firstly demonstrated using the torque envelope as derived in Section 4.3.2. To further improve its performance with feedback signals, an integrated feedback controller design is presented based on the roll angle envelope from Section 4.3.3. For best transient performances under harsh maneuvers, the full LTR envelope from Section 4.3.4 is implemented as constraints in an integrated envelope control architecture.

4.5   FEED-FORWARD TILTING CONTROL

Section 4.3 introduces different implementation of the proposed roll envelope control for an energy-efficient tilting control. To demonstrate how the idea could be adopted in controller designs, this section begins with the most straight-forward use, a feed-forward controller solely for active tilting control.

The major benefits of active tiling control using the feed-forwarded approach are considered as its easy implementation and the proactive control actions. The algorithm is not dependent on roll angle measurements or estimations, and the control action comes proactively which enhances the vehicle roll stability in a highly efficient manner. The implementation of the controller is illustrated in Figure 4.5.

A ramp steering maneuver with vehicle travelling at 60 km/h is simulated. Various roll envelope thresholds LTR_{lim} are adopted, the results of which are shown in Figure 4.6. It can be seen that, for a proper tuned feed-forward controller at low speed, envelope-based active tilting controller can track specified LTR thresholds as envelope boundaries. The more conservative the thresholds are, the more tilting control efforts are required, and vehicle starts to tilt at an earlier stage. The envelope boundary settings can then be used to improves the energy efficiency while still maintain the vehicle stability to a certain level.

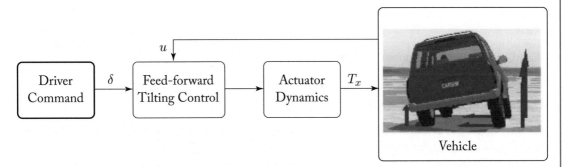

Figure 4.5: Envelope-based feed-forward tilting control.

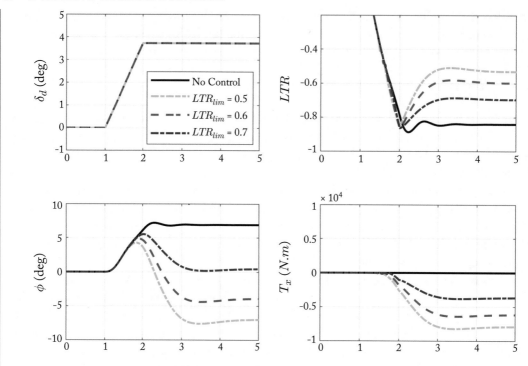

Figure 4.6: Feed-forward roll envelope control.

4.6   INTEGRATED FEEDBACK CONTROL: SMC DESIGN EXAMPLE

4.6.1   MOTIVATION AND OBJECTIVE

Section 4.5 proposes an feed-forward scheme exclusively for tilting control. Although the controller demonstrated the capability to maintain the vehicle roll states within the safe roll envelope, challenges still exist when harsh disturbances are applied to the system and the feed-forward model loses its precision in predicting vehicle behaviors. This section explores an integrated feedback control architecture to tackle this problem.

Apart from that, as inspired by the combined steering-tilting controller designs reviewed in Section 4.2, the suggested control structure also considers vehicle roll dynamics along with the longitudinal and lateral ones in an integrated fashion. Optimal distribution of control efforts as well as systematic consideration of the constraints could be achieved. Tilting moment demand can be reduced, benefiting from this integrated approach compared with separated tilting control schemes, and the lateral stability can also be maintained within the safe region. Some of the previously mentioned challenges in controller designs like flexible architectures and system robustness are addressed with the proposed design.

The integrated controller is aimed at tracking the desired longitudinal speed, while improving the vehicle lateral stability, handling, as well as roll stability. The steady-state roll angle derived in Eq. (4.4) is adopted as the target for the roll envelop control, while control objectives for motions in other directions are described below.

The desired longitudinal u_{des} and lateral speeds v_{des} are derived as

$$u_{des} = u_0 + \int_0^t a_{xd}(\tau)d\tau$$
$$v_{des} = \begin{cases} sign(v)\,\beta_0 u & |v| \geq \beta_0 u \\ v & \text{otherwise,} \end{cases} \qquad (4.8)$$

where u_0 denotes the current vehicle speed; a_{xd} is the acceleration or deceleration command derived from gas/brake pedal; and β_0 represents the threshold of side-slip angles for stability.

The desired yaw rate r_{des} to be tracked for handling improvement is determined from an ideal liner model based on driver's steering input δ, and is also bounded by road friction condition μ:

$$r_{des} = sign(\delta) \cdot \min\left(\left|\frac{u}{l_0 + K_{us}u^2}\delta\right|, \frac{\mu g}{u}\right), \qquad (4.9)$$

where μ denotes the road friction condition to be estimated by algorithms, as suggested in [73, 74].

4.6.2 CONTROL STRUCTURE

A two-layer re-configurable architecture is adopted as illustrated in Figure 4.7.

The top-level controller calculates the generalized forces or moments at vehicle's CoG to track the desired reference states determined in Section 4.6.1. An SMC is adopted due to its robustness against un-modeled dynamics, disturbances, and parameter variations. Considerations of the vehicle's longitudinal, lateral, and roll dynamics in an integrated fashion at this layer helps to utilize the coupling of the dynamics.

The lower-level controller, on the other hand, serves as the control allocator to optimally distribute the desired CoG forces and moments to available actuators considering the vehicle configurations, actuator limitations, as well as tire force capacities. A well-coordinated control action helps to minimize the actuation effort while keeping the vehicle operating in a safe envelope. The re-configurable distribution at this layer improves the system flexibility as well as robustness against actuator failure as will be demonstrated in simulated scenarios.

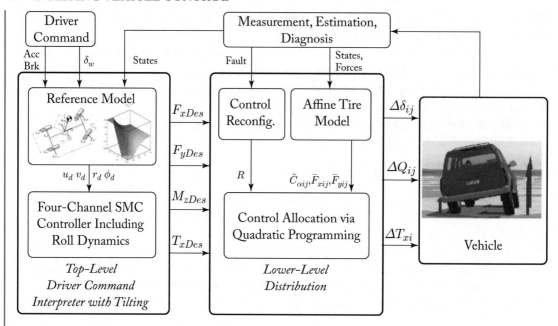

Figure 4.7: Re-configurable SMC controller with integrated control actions [70]. (Image used with permission.)

Top-Level Controller Design

For the top-level tracking control based on the sliding mode approach, sliding surfaces S_i are designed as

$$
\begin{aligned}
S_1 &= u - u_{des} \\
S_2 &= v - v_{des} \\
S_3 &= r - r_{des} \\
S_4 &= \phi - \phi_{des} + \lambda \left(\dot{\phi} - \dot{\phi}_{des} \right),
\end{aligned}
\tag{4.10}
$$

where λ is a positive design parameter for the convergence rate.

For the Lyapunov function candidate $V_i = 1/2 S_i^2$, the attractive sliding surface conditions $\dot{V}_i = -\eta_i \, sign\,(S_i)\, S_i$ can be guaranteed by choosing a sufficiently large η_i. To avoid chattering, the smoothing technique is applied where $sign\,(S_i) \approx \tanh\,(\kappa_i S_i)$.

By plugging vehicle dynamic equations, the virtual forces and moments at vehicle CoG (i.e., F_{xDes}, F_{yDes}, M_{zDes}, T_{xDes}) for performance tracking can be derived as

$$
B \cdot U_{CG} = U_{CGd},
\tag{4.11}
$$

where

$$B = \begin{bmatrix} 1 & 0 & 0 & 0 \\ 0 & 1 & 0 & 0 \\ 0 & 0 & 1 & 0 \\ 0 & \frac{m_s h_s \cos(\phi)}{m} & 0 & 1 \end{bmatrix}, U_{CG} = \begin{bmatrix} F_{xDes} & F_{yDes} & M_{zDes} & T_{xDes} \end{bmatrix}^T$$

$$U_{CGd} = \begin{bmatrix} m\left(\dot{u}_{des} - \eta_1 Sat\left(S_1\right) - vr\right) \\ m\left(\dot{v}_{des} - \eta_2 Sat\left(S_2\right) + ur\right) \\ I_z\left(\dot{r}_{des} - \eta_3 Sat\left(S_3\right)\right) \\ I_x\left(\ddot{\phi}_{des} + \left(\dot{\phi}_{des} - \dot{\phi} - \eta_4 Sat\left(S_4\right)\right)/\lambda\right) + C_\phi \dot{\phi} + K_\phi \phi - m_s g h_s \sin(\phi) \end{bmatrix}.$$

Lower-Level Reconfigurable Control Allocation

At each sampling time, estimated tire forces [75, 76] are used to determine current net forces $\bar{U}_{CG}$ at vehicle CoG. By comparing it with desired virtual forces U_{CG}, the portion should be contributed by control actuation is denoted as ΔU_{CG},

$$U_{CG} = \bar{U}_{CG} + \Delta U_{CG}. \tag{4.12}$$

By adopting the notations from Section 3.3, the control efforts ΔU_{CG} can be rewritten in a re-configurable form as

$$\Delta U_{CG} = LRU, \tag{4.13}$$

where

$$L = \begin{bmatrix} L_\delta \mid L_Q \mid L_{Tx} \end{bmatrix}$$

$$L_Q = \begin{bmatrix} 1/R_{wfl} & 1/R_{wfc} & 1/R_{wfr} & 1/R_{wrl} & 1/R_{wrc} & 1/R_{wrr} \\ 0 & 0 & 0 & 0 & 0 & 0 \\ T_{wl}/2R_{wfl} & T_{wc}/2R_{wfc} & T_{wr}/2R_{wfr} & T_{wl}/2R_{wrl} & T_{wc}/2R_{wrc} & T_{wr}/2R_{wrr} \\ 0 & 0 & 0 & 0 & 0 & 0 \end{bmatrix}$$

$$L_\delta = \begin{bmatrix} 0 & 0 & 0 & 0 & 0 & 0 \\ \tilde{C}_{\alpha fl} & \tilde{C}_{\alpha fc} & \tilde{C}_{\alpha fr} & \tilde{C}_{\alpha rl} & \tilde{C}_{\alpha rc} & \tilde{C}_{\alpha rr} \\ a_f \tilde{C}_{\alpha fl} & a_f \tilde{C}_{\alpha fc} & a_f \tilde{C}_{\alpha fr} & a_r \tilde{C}_{\alpha rl} & a_r \tilde{C}_{\alpha rc} & a_r \tilde{C}_{\alpha rr} \\ 0 & 0 & 0 & 0 & 0 & 0 \end{bmatrix}, L_{Tx} = \begin{bmatrix} 0 \\ 0 \\ 0 \\ 1 \end{bmatrix}.$$

Combining Eqs. (4.11), (4.12), and (4.13), the final distribution equation is simplified as

$$ERU = U_d, \tag{4.14}$$

where $U_d = U_{CGd} - B\bar{U}_{CG}$ is the desired general force need to be distributed an $E = BL$ is the control effectiveness matrix,

$$E = \begin{bmatrix} \frac{1}{R_{wfl}} & \frac{1}{R_{wrr}} & 0 & & 0 & 0 \\ 0 & 0 & \tilde{C}_{\alpha fl} & & \tilde{C}_{\alpha rr} & 0 \\ \frac{T_{wl}}{2R_{wfl}} & \cdots & \frac{T_{wr}}{2R_{wrr}} & a_f \tilde{C}_{\alpha fl} & \cdots & a_r \tilde{C}_{\alpha rr} & 0 \\ 0 & 0 & \frac{m_s h_s \cos(\phi)}{m} \tilde{C}_{\alpha fl} & & \frac{m_s h_s \cos(\phi)}{m} \tilde{C}_{\alpha rr} & 1 \end{bmatrix}.$$

From the structure of the control effectiveness matrix E, it can be seen that torque control is directly effective on longitudinal and yaw dynamics, while active steering control can instantly affect the lateral, yaw, and roll dynamics. Since the tire cornering stiffness is much larger than other constants in the matrix, the steering actuators could be very effective and needs to be properly coordinated with other actuators via the control allocation [70]. The optimal distribution problem is formulated as

$$\min_{U} J = \tfrac{1}{2} \|ERU - U_d\|^2_{W_e} + \tfrac{1}{2} \zeta \|U\|^2_{W_u}$$
$$\text{s.t.} \quad U_{lower} \le U \le U_{upper}, \tag{4.15}$$

where ζ is the scalar weighting parameter and W_e and W_u are the semi-definite weighting matrices for the corresponding terms. The actuator limits are denotes as

$$U_{lower} = \max\left[U_{min}, \bar{U} + \dot{U}_{min} T_s\right]$$
$$U_{upper} = \min\left[U_{max}, \bar{U} + \dot{U}_{max} T_s\right],$$

where $\bar{U}$ is the control efforts being applied at the current sampling time, T_s is the sampling period, $\dot{U}_{max}, \dot{U}_{min}$ denotes the actuation rate limits, and U_{max}, U_{min} denotes the actuation limits, which are defined by the physical actuation limits as well as tire force limits.

Noticing tire forces are dependent on the normal load and the road friction condition as shown in Eq. (3.7), this could be formulated as system constraints as

$$\left(-\bar{F}_{xij} - F^P_{xij}\right) R_{wij} \le \Delta Q_{ij} \le \left(-\bar{F}_{xij} + F^P_{xij}\right) R_{wij}$$
$$\left(-\bar{F}_{yij} - F^P_{yij}\right) \big/ \tilde{C}_{\alpha ij} \le \Delta \delta_{ij} \le \left(-\bar{F}_{yij} + F^P_{yij}\right) \big/ \tilde{C}_{\alpha ij}, \tag{4.16}$$

where $\bar{F}_{xij}$ and $\bar{F}_{yij}$ denote the estimated lateral and longitudinal tire forces and F^P_{xij} and F^P_{yij} denotes the maximum available longitudinal and lateral forces at the current sampling time,

$$F^P_{xij} = F_{xij\,\text{max}} \sqrt{1 - \left(\frac{\bar{F}_{yij}}{F_{yij\,\text{max}}}\right)^2}$$
$$F^P_{yij} = F_{yij\,\text{max}} \sqrt{1 - \left(\frac{\bar{F}_{xij}}{F_{xij\,\text{max}}}\right)^2}.$$

The optimization problem can then be reformulated as a quadratic programming problem, which can be efficiently solved on hardware for run-time applications. Simulation results from [70] are presented below for demonstration.

4.6.3   SIMULATION RESULTS

SMC Case A: Re-Configurable Integrated Vehicle Control

This scenario is designed to demonstrate the re-configurable feature of the integrated control approach. A ramp steering of 0.06 rad is applied to the front axle when the vehicle travels at 60 km/h. Four different actuator configurations are considered:

- Baseline – Four wheel torque vectoring, no roll control:

$$R_{Case1} = diag\left(\begin{bmatrix} 0 & 0 & 0 & 0 & 0 & 0 & 1 & 0 & 1 & 1 & 0 & 1 & 0 \end{bmatrix}\right);$$

- AT + AFS – Active front steering + active tilting:

$$R_{Case2} = diag\left(\begin{bmatrix} 1 & 0 & 1 & 0 & 0 & 0 & 0 & 0 & 0 & 0 & 0 & 0 & 1 \end{bmatrix}\right);$$

- AT + TV – Four wheel torque vectoring + active tilting:

$$R_{Case3} = diag\left(\begin{bmatrix} 0 & 0 & 0 & 0 & 0 & 0 & 1 & 0 & 1 & 1 & 0 & 1 & 1 \end{bmatrix}\right);$$

- AT + AFS + TV – Active front steering + four wheel torque vectoring + tilting:

$$R_{Case4} = diag\left(\begin{bmatrix} 1 & 0 & 1 & 0 & 0 & 0 & 1 & 0 & 1 & 1 & 0 & 1 & 1 \end{bmatrix}\right).$$

Since the roll dynamics is very dependent on the lateral acceleration disturbances, the vehicle with torque vectoring (Case 1) for yaw rate tracking is chosen as the baseline for a fair comparison. The active steering is applied to both front wheels, and a roll envelope boundary with $LTR_{lim} = 0.5$ is adopted for the demonstration. Results in Figure 4.8 show that the proposed envelope controller with different actuation systems successfully mitigate the rollover danger by reducing LTR index from 0.7 to 0.5.

Although similar roll and lateral stability performances can be achieved with different actuators using the proposed integrated control scheme, the control efforts are different. Case 3 (AT+TV) requires most tilting torques, as the desired yaw rate is tracked by torque vectoring, and most of the desired roll moment is generated by tilting actuators.

Case 2 (AT+AFS) adopts active steering actuators. As shown in the control effectiveness matrix, apart from the yaw motion tracking, lateral tire forces generated by active steering could directly help the rollover mitigation. A compromise can then be made to balance two control targets. In this simulation, the weight is tuned to give a priority to the rollover mitigation [70]. The handling performance evaluated by the yaw rate tracking is decreased. Consequently, the control effort exerted by the tilting mechanism also gets reduced.

When all actuators are available as shown in Case 4 (AT+AFS+TV), the vehicle performances can be further improved. The desired yaw rate can be better tracked by utilizing both torque vectoring and active steering when roll envelope control is not activated, and control efforts could be minimized between actuators. However, when the roll mitigation control is required, the active steering could be tuned to help reducing the active tilting moment, while the vehicle handling performance is not compromised. The desired yaw rate, as shown in the figure, could still be maintained by the torque vectoring thanks to the integrated formulation.

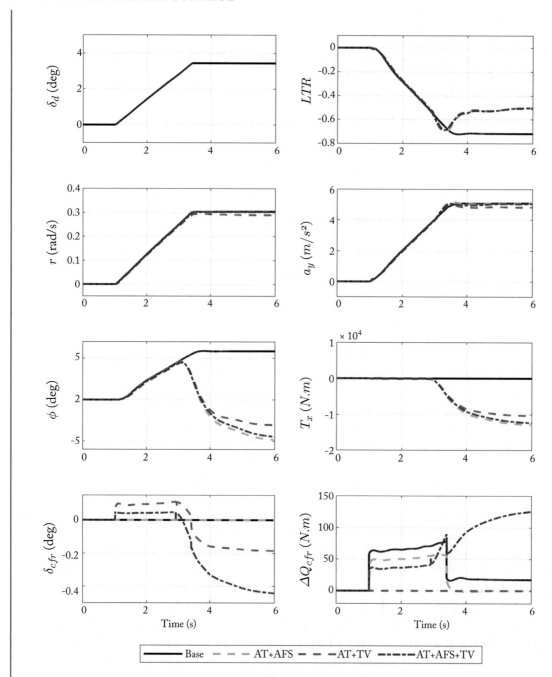

Figure 4.8: Various actuator configurations with the roll envelope-based control.

SMC Case B: Robustness to Tilting Actuator Failure

To demonstrate the robustness of the proposed integrated control architecture, the vehicle configuration with the following Case 5 is adopted.

- Case 5 – Same configuration as Case 4, but tilting actuator could fail during runtime:

$$R_{Case5} = diag\left(\begin{bmatrix} 1 & 0 & 1 & 0 & 0 & 0 & 1 & 0 & 1 & 1 & 0 & 1 & R_{Tx}(t) \end{bmatrix}\right),$$

where $R_{Tx}(t)$ is the diagnosis of the actuator health condition which can be determined by algorithms suggested in [21, 39, 63].

A failure is triggered from time $t = 4$ s with tilting torque output drops to zero. Two scenarios are compared. One assumes there is no flexibility in the control architecture to handle such faults, and the controller is not notified about the system change and still expects the actuator to exert the same amount of torque in its normal state as Case 4; in the other scenario, a fault-diagnostic system could capture the failure as Case 5 and reformulate the optimal distributions online. Considering the driver might not immediately realize the system faults, vehicle speed is assumed to remain unchanged at 60 km/h for both scenarios. Results are shown in Figure 4.9.

The baseline case (Case 1) drives the vehicle with a fishhook maneuver. Since no active tilting control is involved, the LTR index could reach more than 0.7. With the demonstrated integrated control of multiple actuators, vehicle LTR indices of both can be significantly reduced when there is no system faults ($t < 4$ s). After tilting actuator fault is triggered, the vehicle behavior without control reconstruction will put the vehicle into the high rollover danger again as the baseline case [70]. However, with the proposed re-configurable approach, control efforts can be re-distributed online to react to system changes. Vehicle rollover stability could still be maintained at a safe level, with minimum impacts to motions in other directions.

4.7   HOLISTIC ENVELOPE CONTROL: MPC DESIGN EXAMPLE

Previous sections represent two different approaches to tackle the tilting control problem: a separate tilting control and an integrated stability control with active tilting formulations. The idea of the roll envelope derived from the steady-state assumptions are adopted in both architectures. Good performances have been demonstrated, however ignoring the transient roll dynamics still limits their performances.

This section explores how the full roll envelope as constraints can be incorporated in the control system design, which leads to enhanced performances. More specifically, the non-minimum phase problem with active roll motion control is discussed, and the model predictive controller is presented to tackle the control problem. For clarity, the envelope-based MPC exclusively for roll motion control is presented first, followed by the integrated stability envelope approach.

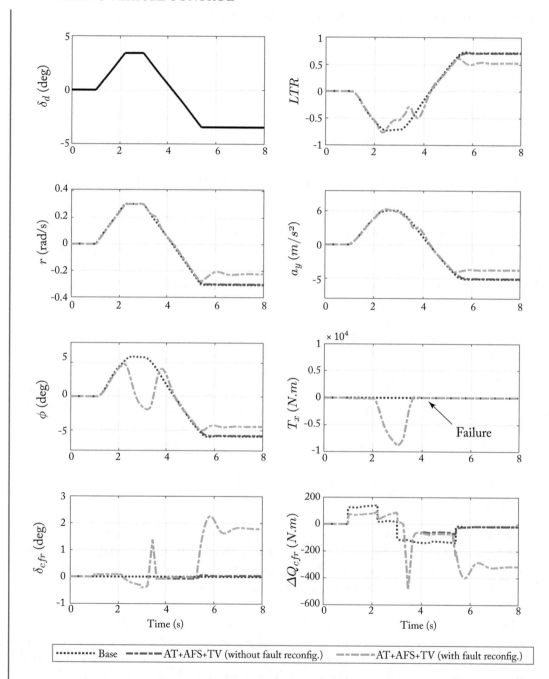

Figure 4.9: Roll envelope-based control with actuator failure.

4.7.1 NON-MINIMUM PHASE IN ACTIVE TILTING CONTROL

The non-minimum phase system are known as any dynamical system whose transfer function has RHS zeros [77]. Before any detailed derivations are given, an intuitive way to look at this is through the phase plot of the roll envelope introduced in Section 4.3.4.

Previous envelope plots are drawn for cases with small or no disturbances, in which no control actions are taken. However, when disturbances are estimated to cause rollover dangers, active tilting control should be applied for rollover mitigation. As shown in Eq. (3.14), the rollover index is dependent on motion states as well as control inputs. When tilting moment is applied, the envelope starts to shift instantly as described by the algebraic equation. This could cause problems, as demonstrated in Figure 4.10.

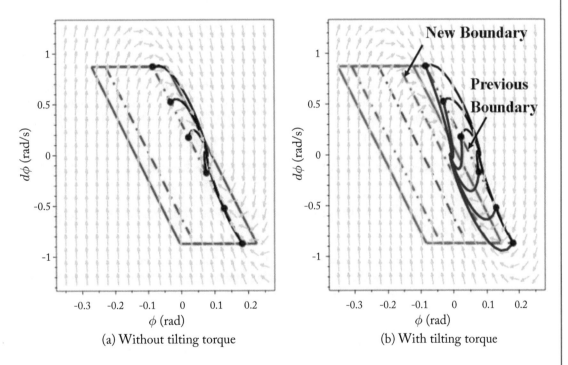

(a) Without tilting torque

(b) With tilting torque

Figure 4.10: Effects of tilting torque on the roll envelope.

The uncontrolled roll dynamics under high lateral acceleration disturbances is shown in Figure 4.10a. As before, two sets of LTR boundaries are plotted. Given initial states (denoted by dots) on the boundary, they will converge to the equilibrium point which is outside the safe envelope. As a comparison, the dynamics with a constant tilting torque is illustrated in Figure 4.10b. Starting with same initial conditions, the new state trajectories in blue are now converging to the inner side of the original boundary. This definitely helps the roll stability of NTVs in steady-state, however, transient performances could be deteriorated due to the boundary shift with the applied torque. The new LTR envelope boundaries considering the effect of tilting moments are plotted in pink. Since it takes time for roll states to converge to the new equilibrium while the LTR envelope changes instantly after the torque is applied, the LTR index gets even worse during the transient phase.

The RHS zero, according to the definition of non-minimum phase systems, can be explicitly shown by combining Eqs. (3.9) and (3.14). The transfer function from tilting moment to LTR index is written as

$$H(s) = C_r(sI - A_r)^{-1}B_r + D_r$$
$$= \frac{2(I_x s^2 - m_s h_s g)}{(m_s + m_u)T_w g (I_x s^2 + C_\phi s + K_\phi - m_s h_s g)}. \tag{4.17}$$

The zeros of the above transfer function can be solved as

$$s = \pm\sqrt{m_s h_s g / I_x}, \tag{4.18}$$

which indicates there always exists an RHS zero. A fundamental limitation between response time and system overshoot exists according to [78]. Higher tilting moment helps to change the roll dynamics quicker at the cost of more severe LTR index overshoot. From a physical perspective, this can be explained by referring to the force analysis in Figure 3.2, as

1. to balance the lateral acceleration (a_y) with gravity, a clockwise roll angle in steady state is desired;

2. the tilting torque applied to the sprung mass needs to be in the clockwise direction, so as the roll acceleration;

3. treating the sprung and un-sprung mass as a whole, tilting moment as an internal force does not change the load distribution directly, but the inertia moment due to the roll acceleration makes the LTR index worse; and

4. the fundamental limitation exists between the steady-state target (achieve desired roll angle as soon) and the transient objective (small roll accelerations to avoid LTR overshoot).

A quick reaction time is definitely desired for emergent rollover mitigation, while the transient overshoot should be carefully monitored and bounded in the controller design. To address this, an optimal controller based on the vehicle roll model is proposed to compromise between these conflicting targets. The prediction capability in the controller is considered quite beneficial since it can foresee the arriving of the envelope violation and applies the tilting moment at an early stage to minimize the overshoot.

4.7.2   ROLL CONSTRAINT ENVELOPE WITH MPC

To apply the above-mentioned safe envelope as state constraints in the controller design, the model predictive control approach is adopted due to its straightforward implementation of the constraints regarding the states and inputs in the form of Eq. (4.7). The predictive feature of the controller also enables it to foresee the violation of the envelope and react at an early stage.

As shown in Section 4.7.1, a strict LTR constraint formulation could lead to an infeasible problem since the use of tilting actuators always comes with the side-effect of decreased stability in transient. However, if such LTR overshoot can be bounded, perturbation of the states outside of the boundary allows the system to react more quickly for emergency conditions [70]. The slack variable is thus introduced for a soft constraint formulation by allowing constraint violations but minimizing them in the cost function against other control objectives. For implementations with the MPC scheme, the following formulation for the slack variables in the prediction horizon $S^{(k)}$ is adopted:

$$\min_{U,S} : \quad J = \sum_{k=1}^{N} \left\| S^{(k)} \right\|_p^Q$$

$$\text{s.t.} : \quad X^{(k+1)} = AX^{(k)} + BU^{(k)} + EW^{(0)} \qquad (4.19)$$

$$M X^{(k)} + N U^{(k)} \leq L + S^{(k)}$$

$$S^{(k)} \geq 0,$$

where $\| \cdot \|_p^Q$ denotes the norm-p of slack variables weighted by matrix Q, Matrix A, B, E represents general system matrices, while M, N, L denotes the roll envelope as derived in Eq. (4.7).

The cost of the slack variable should be minimized to avoid constraint violation in both the magnitude as well as the duration, which are competing objectives for non-minimum phase systems as described. The l_∞ norm captures the worst-case constraint violation but does not consider the duration of the violation time. It is also reported to have non-intuitive parameter tuning issues as well as poor closed-loop performance [79]. The l_2 norm is chosen for the slack variables in this research. Compared with the l_1 norm which linearly penalizes the constraint violation, the l_2 norm better captures the demand to maintain the vehicle away from the "hard boundary" for vehicle rollover conditions ($LTR = \pm 1$). It should be mentioned that the adoption of l_1 or l_2 norm increases the computational burden, but it has been shown to be real-time tractable [80] with the current hardware limitations.

With control objectives to maintain the vehicle within the safe roll envelope as well as minimizing control efforts, an optimal receding horizon control is formulated as

$$\min_{U_r, S_{ro}} : \quad J = \sum_{k=1}^{N} \left\| U_r^{(k)} \right\|_2^{R_{ro}} + \sum_{k=1}^{N} \left\| S_{ro}^{(k)} \right\|_2^{Q_{ro}}$$

$$\text{s.t.} : \quad X_r^{(k+1)} = A_{rd} X_r^{(k)} + B_{rd} U_r^{(k)} + E_{rd} W_r^{(0)}$$

$$M_{ro} X^{(k)} + N_{ro} U^{(k)} \leq L_{ro} + S_{ro}^{(k)} \tag{4.20}$$

$$0 \leq S_{ro}^{(k)} \leq S_{ro,\max}$$

$$\left| U_r^{(k)} \right| \leq U_{\max}$$

$$\left| U_r^{(k)} - U_r^{(k-1)} \right| \leq U_{slew,\max},$$

where $S_{ro}^{(k)} \in \mathbb{R}^4$ denotes slack variables for the roll envelope at the $k-th$ prediction step. The disturbance is assumed to remain unchanged during the control horizon as $W^{(0)}$. A_{rd}, B_{rd}, E_{rd} are the discretized matrices of A_r, B_r, E_r derived from the roll model. The total cost to be minimized is the l_2 norm of the control effort and slack variables, which are weighted by R_{ro} and Q_{ro}, respectively.

The choice of the prediction horizon, as detailed in [70], should be long enough to see the steady-state performance enhancement along with the transient degradation for an optimal decision making. A ramp acceleration disturbance of $0.5g$ is applied to the system, and the roll envelope is set as $LTR_{\lim} = 0.5$. MPC results with various prediction horizon settings are shown in Figure 4.11. For a small prediction horizon as 0.5 s (i.e., $N = 10$, $\Delta T = 50$ ms), the resultant tilting moment T_x is in the wrong direction and makes LTR even worse. Results for the optimized torque start to converge after the prediction horizon reaches 1 s. For computational efficiency, a prediction horizon of 1 s is adopted for the control implementations.

It should also be noted that the LTR index overshoot is still noticeable as shown in Figure 4.11 with the active tilting control only. To further improve the vehicle stability, other on-board control efforts should be incorporated, and an integrated approach is introduced in the next section.

4.7.3   INTEGRATED STABILITY ENVELOPE

For lateral stability, a handling envelope has been proposed by researchers [81, 82]. The thresholds for yaw rate ($r_{\max}$) along with side-slip angles at rear wheels ($\alpha_{r,sat}$) are used to formulate the envelope for lateral stability control. At each discrete control step k, this leads to

$$M_{sh} X^{(k)} \leq L_{sh}, \tag{4.21}$$

where

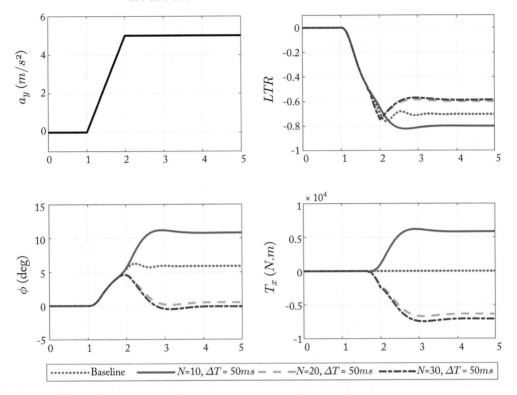

Figure 4.11: Roll envelope control with different prediction horizons.

$$M_{sh} = \begin{bmatrix} +1 & -b/u \\ -1 & +b/u \\ 0 & +1 \\ 0 & -1 \end{bmatrix} \quad O_{4\times2}, \quad L_{sh} = \begin{bmatrix} \alpha_{r,sat} \\ \alpha_{r,sat} \\ r_{max} \\ r_{max} \end{bmatrix}.$$

The handling envelope can be visualized in the phase plane as the boundaries shown in Figure 4.12a. The phase plane is defined by states of lateral speed v and yaw rate r. The horizontal boundaries stand for the yaw rate limit (r_{max}), while the sloped lines represent the constraints for rear tire slip angle ($\alpha_{r,sat}$). For a less harsh maneuver (shown in blue) when the state trajectory stays within the handling envelope, no control intervention is required. Active safety control will be applied only when the system foresees the approaching of envelope violations.

By combining the handling envelope with the proposed roll envelope, an integrated envelope approach can be formulated, as shown in the Figure 4.12. Vehicle status is considered safe when both lateral and roll envelope constraints are satisfied. Compared with the conventional tracking approach for stability control, the proposed integrated envelope-based controller

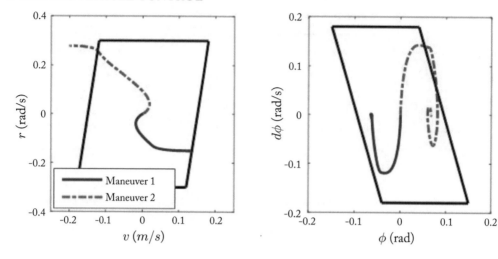

Figure 4.12: Integrated envelope approach for vehicle stability.

applies the control effort only when the predicted vehicle states are leaving the safe envelopes, which reduces the control intervention while still maintain the desired level of stability. The fact that lateral control efforts like active steering, torque vectoring and differential braking can help the rollover mitigation could also be implicitly incorporated when formulating the motion control problem in both directions as a whole.

For implementation, the integrated re-configurable model developed in Section 3.3 is adopted. Based on that, an integrated vehicle controller with various actuator configurations is demonstrated using the proposed envelope control scheme for both lateral and roll stability. The controller performance, as well as robustness, is shown to be further improved with the proposed integrated approach. The previously suggested roll envelope control Eq. (4.20) could be extended for the integrated envelope control as

$$\min_{U, S_{ro}, S_{sh}} : J = \tfrac{1}{2} \sum_{k=1}^{N} \left\| U^{(k)} \right\|_2^{R_U} + \tfrac{1}{2} \sum_{k=1}^{N} \left\| r^{(k)} - r_{des} \right\|_2^{R_X} + \dots$$
$$\tfrac{1}{2} \sum_{k=1}^{N} \left\| S_{ro}^{(k)} \right\|_2^{Q_{ro}} + \tfrac{1}{2} \sum_{k=1}^{N} \left\| S_{sh}^{(k)} \right\|_2^{Q_{sh}}$$

$$\text{s.t. :} \qquad X^{(k+1)} = A_d X^{(k)} + B_d U^{(k)} + E_d W^{(0)}$$
$$M_{sh} X^{(k)} \leq L_{sh} + S_{sh}^{(k)}$$
$$S_{sh}^{(k)} \geq 0$$

(4.22)

$$M_{ro}X^{(k)} + N_{ro}U^{(k)} \le L_{ro} + S_{ro}^{(k)}$$
$$0 \le S_{ro}^{(k)} \le S_{ro,\max}$$
$$\left|U^{(k)}\right| \le U_{\max}$$
$$\left|U^{(k)} - U^{(k-1)}\right| \le U_{slew,\max},$$

where A_d, B_d, and E_d are discretized system matrices. $S_{sh}^{(k)} \in \mathbb{R}^4$ and $S_{ro}^{(k)} \in \mathbb{R}^4$ denotes handling and roll envelope slack variables at the prediction step k. The disturbance is assumed to remain unchanged during the control horizon as $W^{(0)}$. The total cost to be minimized is composed of control efforts, yaw rate tracking errors, and the slack variable for both lateral and roll envelopes, which are weighted by R_U, R_X, Q_{ro}, and Q_{sh}, respectively. The desired yaw rate to be tracked is the same as shown in Eq. (4.9).

Using the batch formulation, state trajectories in the prediction horizon, by adopting the plant model, can be written as

$$\bar{X} = S_X X^{(0)} + S_{U0}U^{(0)} + S_{W0}W^{(0)} + S_U \bar{U} + S_W \bar{W}, \tag{4.23}$$

where

$$\bar{X} = \begin{bmatrix} X^{(1)} \\ X^{(2)} \\ \vdots \\ \vdots \\ X^{(N)} \end{bmatrix}, \bar{U} = \begin{bmatrix} U^{(1)} \\ U^{(2)} \\ \vdots \\ \vdots \\ U^{(N)} \end{bmatrix}, \bar{W} = \begin{bmatrix} W^{(1)} \\ W^{(2)} \\ \vdots \\ \vdots \\ W^{(N)} \end{bmatrix}$$

$$S_X = \begin{bmatrix} A_d^1 \\ A_d^2 \\ \vdots \\ \vdots \\ A_d^N \end{bmatrix}, S_{U0} = \begin{bmatrix} B_d^1 \\ A_d B_d \\ \vdots \\ \vdots \\ A_d^{N-1} B_d \end{bmatrix}, S_{W0} = \begin{bmatrix} E_d^1 \\ A_d E_d \\ \vdots \\ \vdots \\ A_d^{N-1} E_d \end{bmatrix}$$

$$S_U = \begin{bmatrix} O & \cdots & \cdots & \cdots & O \\ B_d & O & \cdots & \cdots & O \\ A_d B_d & B_d & O & \cdots & \vdots \\ \vdots & \vdots & \ddots & \ddots & \vdots \\ A_d^{N-2} B_d & A_d^{N-3} B_d & \cdots & B_d & O \end{bmatrix}$$

$$S_W = \begin{bmatrix} O & \cdots & \cdots & \cdots & O \\ E_d & O & \cdots & \cdots & O \\ A_d E_d & E_d & O & \cdots & \vdots \\ \vdots & \vdots & \ddots & \ddots & \vdots \\ A_d^{N-2} E_d & A_d^{N-3} E_d & \cdots & E_d & O \end{bmatrix}.$$

The envelope constraints can also be written with the batch form. For example, the roll envelope as

$$M_{RO}\bar{X} + N_{RO}\bar{U} \leq L_{RO} + \bar{S}_{ro}, \tag{4.24}$$

where

$$M_{RO} = BlockDiag\left(\begin{bmatrix} M_{ro} & M_{ro} & \cdots & \cdots & M_{ro} \end{bmatrix}\right)$$
$$N_{RO} = BlockDiag\left(\begin{bmatrix} N_{ro} & N_{ro} & \cdots & \cdots & N_{ro} \end{bmatrix}\right)$$
$$L_{RO} = \begin{bmatrix} L_{ro} & L_{ro} & \cdots & \cdots & L_{ro} \end{bmatrix}^T$$
$$\bar{S}_{RO} = \begin{bmatrix} S_{ro}^{(1)} & S_{ro}^{(2)} & \cdots & \cdots & S_{ro}^{(N)} \end{bmatrix}^T.$$

Combining Eqs. (4.23) and (4.24) gives the design variables in the linear constraint form

$$A_{RO}\eta \leq b_{RO}, \tag{4.25}$$

where

$$\eta = \begin{bmatrix} \bar{U} & \bar{S}_{ro} & \bar{S}_{sh} \end{bmatrix}^T$$
$$A_{RO} = BlockDiag\left(\begin{bmatrix} M_{RO}S_U + N_{RO} & -I & O \end{bmatrix}\right)$$
$$b_{RO} = L_{RO} - M_{RO}\left(S_X X^{(0)} + S_{U0}U^{(0)} + S_{W0}W^{(0)} + S_W \bar{W}\right).$$

Similarly, the handling envelope can be written as

$$A_{SH}\eta \leq b_{SH} \tag{4.26}$$

with

$$A_{SH} = BlockDiag\left(\begin{bmatrix} M_{SH}S_U & O & -I \end{bmatrix}\right)$$
$$b_{SH} = L_{SH} - M_{SH}\left(S_X X^{(0)} + S_{U0}U^{(0)} + S_{W0}W^{(0)} + S_W \bar{W}\right).$$

Equations (4.25) and (4.26), along with the constraints on the design variables and their slew rate, form the linear constraint in the quadratic programming problem. The objective function in Eq. (4.22) can be rewritten by considering Eq. (4.23) as

$$J = \frac{1}{2}\eta^T H \eta + f\eta + Const., \tag{4.27}$$

where

$$H = BlockDiag \left(\begin{bmatrix} \left(R_U + S_U^T R_X S_U \right) & Q_{ro} & Q_{sh} \end{bmatrix} \right)$$

$$f = \begin{bmatrix} \left(S_X X^{(0)} + S_{U0} U^{(0)} + S_{W0} W^{(0)} + S_W \bar{W} - \bar{X}_{des} \right)^T R_X S_U & O & O \end{bmatrix}^T.$$

The quadratic programming can then be written in the form as

$$\min_{\eta} : \quad J = \frac{1}{2}\eta^T H \eta + f\eta$$

$$\text{s.t.} : \quad \begin{bmatrix} A_{RO} \\ A_{SH} \end{bmatrix} \eta \le \begin{bmatrix} b_{RO} \\ b_{SH} \end{bmatrix} \tag{4.28}$$

$$\eta_{\min} \le \eta \le \eta_{\max}$$

$$\left| \eta^{(k)} - \eta^{(k-1)} \right| \le \eta_{slew,\max}.$$

4.7.4 SIMULATION RESULTS

The system diagram for the simulation is shown in Figure 4.13. The re-configurable model is

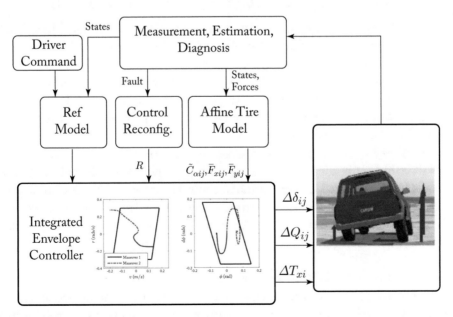

Figure 4.13: Re-configurable integrated envelope control.

again adopted to compare the envelope control performances with various actuator configurations. The robustness of the proposed approach is also shown via the scenario considering the tilting actuator failure.

MPC Case A: Re-Configurable Integrated Vehicle Control

In this scenario, the re-configurable feature of the suggested integrated envelope controller is demonstrated. A ramp steering signal shown in Figure 4.14 is applied to both front wheels at $t = 1.0$ s with vehicle longitudinal speed maintained at 60 km/h. The following four actuator configurations are considered:

- Baseline – Four wheel torque vectoring, no roll control:

$$R_{Case1} = diag\left(\begin{bmatrix} 0 & 0 & 0 & 0 & 0 & 0 & 1 & 0 & 1 & 1 & 0 & 1 & 0 \end{bmatrix}\right);$$

- AT + AFS – Active front steering + active tilting:

$$R_{Case2} = diag\left(\begin{bmatrix} 1 & 0 & 1 & 0 & 0 & 0 & 0 & 0 & 0 & 0 & 0 & 0 & 1 \end{bmatrix}\right);$$

- AT + TV – Four wheel torque vectoring + active tilting:

$$R_{Case3} = diag\left(\begin{bmatrix} 0 & 0 & 0 & 0 & 0 & 0 & 1 & 0 & 1 & 1 & 0 & 1 & 1 \end{bmatrix}\right);$$

- AT + AFS + TV – Active front steering + four wheel torque vectoring + tilting:

$$R_{Case4} = diag\left(\begin{bmatrix} 1 & 0 & 1 & 0 & 0 & 1 & 0 & 1 & 1 & 0 & 1 & 1 \end{bmatrix}\right).$$

It should be noted that, compared with the steering maneuver adopted in simulations of SMC Case A, the proposed envelope controller with full constraints formulation is demonstrated with harsher steering maneuvers. The rollover threshold is still chosen as $LTR_{lim} = 0.5$. As can be seen from the results shown in Figure 4.14, all configurations can mitigate the rollover danger by decreasing the LTR index from more than 0.8 down to the desired threshold 0.5. The vehicle also gets stabilized much quicker, compared with the previous SMC designs, while the overshoot of LTR is maintained at a safe level.

Case 3 (AT+TV) also suggests that, apart from existing STC or SDTC controller designs for NTV which only adopts the steering actuators, more generalized lateral control approaches could be incorporated via the integrated control framework. When all actuators Case 4 (AT+AFS+TV) are available, much smoother performance can be obtained. More counter steering is applied to ease the tilting, while the desired yaw rate is still maintained by the torque vectoring as shown in the figure.

MPC Case B: Robustness to Tilting Actuator Failure

Similar to the SMC design, the fault-tolerance feature could also be demonstrated with the integrated envelope architecture. The vehicle with redundant actuators, as shown in Case 4 (AT+AFS+TV), is adopted. Actuator failure is simulated by changing $R_{Tx}(t)$ in Case 5 during

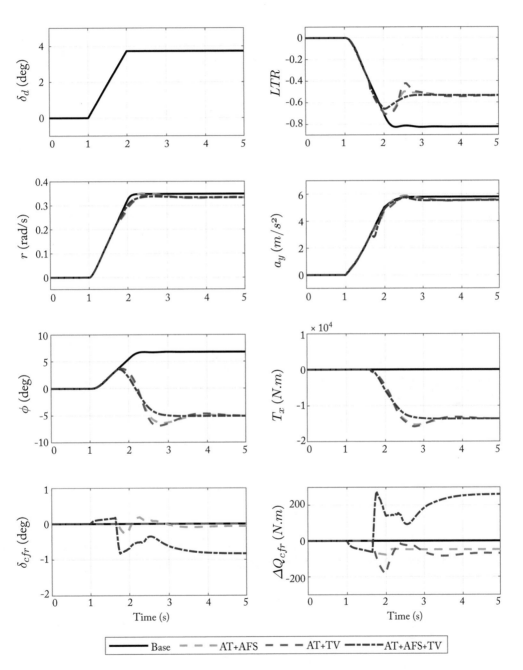

Figure 4.14: Various actuator configurations with the integrated envelope control.

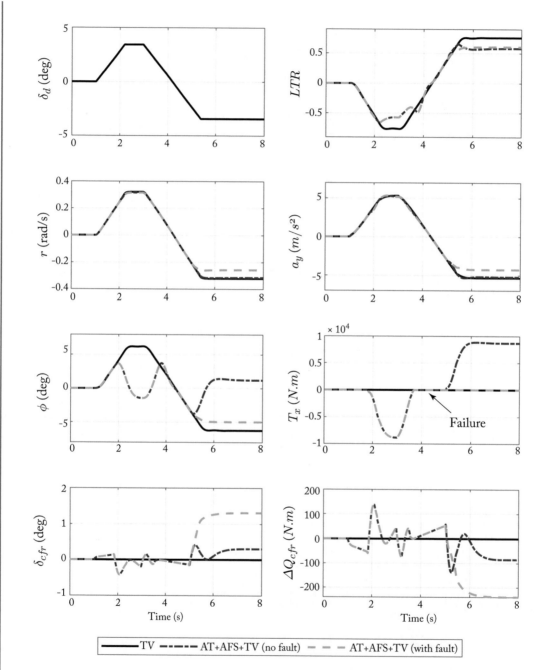

Figure 4.15: Integrated envelope control with actuator failure.

run-time to indicate the health status of the tilting actuators. Output from the tilting actuators are expected to drop to zero after the failure is injected as shown for T_x after $t = 4$ s in
Figure 4.15.

- Case 5 – Same configuration as Case 4, but with tilting actuator failure during runtime:

$$R_{Case5} = diag \left(\begin{bmatrix} 1 & 0 & 1 & 0 & 0 & 0 & 1 & 0 & 1 & 1 & 0 & 1 & R_{Tx}(t) \end{bmatrix} \right),$$

where $R_{Tx}(t)$ is the diagnosis of the actuator health condition which can be determined by algorithms suggested in [21, 39, 63].

Two scenarios are compared for the fault-tolerant control. One assumes there is no fault
and the system behaves as the normal Case 4, while the other with injected faults and a fault-
diagnostic system is assumed to capture the error and update the internal model during run-time
(i.e., Case 5) for control re-distributions. A torque vectoring configuration with no tilting control
is adopted as the baseline.

A fishhook maneuver is applied when the vehicle travels at 60 km/h. Before the fault is
triggered, both controller behaves the same and vehicle rollover danger can be greatly reduced
as shown in the figure. After the injected tilting actuator fault, the controller with runtime re-
configurability behaves more robust. As the shown in the results, the steering control will be
more active after the tilting actuator failure to compensate for the lost of control moments,
while torque vectoring actuators will function to keep the yaw rate changes to a minimum. As a
consequence of this redistribution, the vehicle will behave more under-steer compared with the
normal case, but the rollover index is successfully maintained as a normal vehicle, which gives
the driver enough time to handle the situation.

CHAPTER 5

Conclusions

This book investigated the active tilting system for narrow tilting vehicles (NTVs) from different angles including concept, mechanism, dynamics, and control.

A review of issues in current urban transportation revealed the insufficient use of conventional vehicles, and led to narrow urban vehicles for congested urban areas. Congestion, parking, and pollution issues can be greatly relieved by the reduced footprint and weight of the vehicle. Active tilting, which mimics cyclist maneuvers during cornering for roll stability enhancement, is considered vital for successful NTV designs.

Various tilting mechanisms were examined with focus on their space requirements, complexity, and effectiveness against vehicle rollover. An integrated suspension tilting system (ISTS) was suggested as an alternative method to synchronize the vehicle tilting motion in replacement for mechanical linkages. The wheel synchronization problem which used to be enforced by mechanical constraints were tackled by the coordination of hydraulic pumps embedded in active hydraulic interconnected suspensions (HIS).

The rest of the book dealt with the modeling and control of active tilting actuators for rollover mitigation. Full vehicle dynamics as well as stability measures were discussed. With the derived lateral load transfer (LTR) index from modeling section, an envelope-based controller was suggested for NTVs to alleviate the energy consumption on tilting actuators to achieve a more energy-efficient personal carrier platform. Various implementations of the proposed envelope approach was demonstrated via controller architectures with different complexities. Compared with existing tilting strategies which try to eliminate the lateral load transfer during vehicle cornering, the proposed envelope approach was featured by less frequent activation of tilting actuators and was shown to save the titling effort especially during non-harsh urban driving scenarios.

Bibliography

[1] K. M. Kockelman and Y. Zhao, Behavioral distinctions: The use of light-duty trucks and passanger cars, *Journal of Transportation and Statistics*, vol. 3, no. 3, December 2000. https://trid.trb.org/view/685651 1

[2] R. Hibbard and D. Karnopp, Twenty first century transportation system solutions—a new type of small, relatively tall and narrow active tilting commuter vehicle, *Vehicle System Dynamics*, vol. 25, no. 5, pp. 321–347, May 1996. http://www.tandfonline.com/doi/abs/10.1080/00423119608968970 DOI: 10.1080/00423119608968970. 1, 3, 4, 31, 32

[3] J. C. Huston, B. J. Graves, and D. B. Johnson, Three wheeled vehicle dynamics, *SAE Technical Paper*, February 1982. http://papers.sae.org/820139/ DOI: 10.4271/820139. 1

[4] F. Will, J. N. Davdison, P. Couchman, and D. Bednall, Tomorrow's car—for today's people: Can tilting three wheeled vehicles be a solution for the problems of today and the future? *SAE Technical Paper*, Tech. Rep., 2011. DOI: 10.4271/2011-28-0001. 1

[5] A. Dosovitskiy, G. Ros, F. Codevilla, A. Lopez, and V. Koltun, Carla: An open urban driving simulator, *Proc. of the 1st Annual Conference on Robot Learning*, pp. 1–16, 2017. 2

[6] P. L. Boyd, NHTSA's NCAP rollover resistance rating system, *19th International Technical Conference on the Enhanced Safety of Vehicles (ESV)*, June 2005. https://trid.trb.org/view/811348 2

[7] S. Kidane, L. Alexander, R. Rajamani, P. Starr, and M. Donath, A fundamental investigation of tilt control systems for narrow commuter vehicles, *Vehicle System Dynamics*, vol. 46, no. 4, pp. 295–322, April 2008. http://www.tandfonline.com/doi/abs/10.1080/00423110701352987 DOI: 10.1080/00423110701352987. 3, 32, 33

[8] M. I. Barker, Chassis design and dynamics of a tilting three-wheeled vehicle, Ph.D., University of Bath, 2006. http://ethos.bl.uk/OrderDetails.do?uin=uk.bl.ethos.432834 4, 7

[9] N. Amati, A. Festini, L. Pelizza, and A. Tonoli, Dynamic modelling and experimental validation of three wheeled tilting vehicles, *Vehicle System Dynamics*, vol. 49, no. 6, pp. 889–914, June 2011. http://www.tandfonline.com/doi/abs/10.1080/00423114.2010.503277 DOI: 10.1080/00423114.2010.503277. 8, 10

[10] C. Tang, A. Goodarzi, and A. Khajepour, A novel integrated suspension tilting system for narrow urban vehicles, *Proc. of the Institution of Mechanical Engineers, Part D: Journal of Automobile Engineering*, p. 095440701773827, November 2017. `http://journals.sagepub.com/doi/10.1177/0954407017738274` DOI: 10.1177/0954407017738274. 9, 11, 12, 13

[11] J.-C. Chiou and C.-L. Chen, Modeling and verification of a diamond-shape narrow-tilting vehicle, *IEEE/ASME Transactions on Mechatronics*, vol. 13, no. 6, pp. 678–691, 2008. DOI: 10.1109/tmech.2008.2004769. 4, 10, 17

[12] marsMediaSite, The curve master—F 300 life jet. `http://media.daimler.com/marsMediaSite/en/instance/ko/The-curve-master--F-300-Life-Jet.xhtml?oid=9272608` 4

[13] C. Technology, The unique self-balancing vehicle technology hold by Carver Technology. `http://www.carver-technology.com/` 5

[14] S. Maakaroun, W. Khalil, M. Gautier, and P. Chevrel, Modeling and simulating a narrow tilting car using robotics formalism, *IEEE Transactions on Intelligent Transportation Systems*, vol. 15, no. 3, June 2014. DOI: 10.1109/tits.2013.2293125. 5, 9, 10

[15] T. M. Corporation, Toyota global site | personal mobility | Toyota i-road. `http://www.toyota-global.com/innovation/personal_mobility/i-road/index.html` 5

[16] A. C. Zolotas, Advanced control strategies for tilting trains, thesis, ©Argyrios C. Zolotas, 2002. `https://dspace.lboro.ac.uk/dspace-jspui/handle/2134/4279` 7

[17] J. J. H. Berote Dynamics and control of a tilting three wheeled vehicle, Ph.D. thesis, University of Bath, 2010. `http://opus.bath.ac.uk/24682/` 7

[18] A. Van Poelgeest, The dynamics and control of a three-wheeled tilting vehicle, Ph.D. thesis, University of Bath, 2011. `http://opus.bath.ac.uk/24680/` 30

[19] J. Robertson, Active control of narrow tilting vehicle dynamics, Ph.D. thesis, University of Bath, February 2014. `http://opus.bath.ac.uk/43311/`

[20] H. Alipour, M. B. B. Sharifian, and M. Sabahi, A modified integral sliding mode control to lateral stabilisation of 4-wheel independent drive electric vehicles, *Vehicle System Dynamics*, vol. 52, no. 12, pp. 1584–1606, December 2014. `http://dx.doi.org/10.1080/00423114.2014.951661` DOI: 10.1080/00423114.2014.951661. 7, 30

[21] H. K. C. R. van den Brink, DVC—The banking technology driving the carver vehicle class, *AVEC*, Arnhem, The Netherlands, 2004. `http://www.carver-technology.com/media.html` 7, 10, 33, 49, 63

[22] J. Edelmann, M. Plöchl, and P. Lugner, Modeling and analysis of the dynamics of a tilting three-wheeled vehicle, *Multibody System Dynamics*, vol. 26, no. 4, pp. 469–487, December 2011. `https://link.springer.com/article/10.1007/s11044-011-9258-7` DOI: 10.1007/s11044-011-9258-7. 8, 9, 17

[23] S. Kidane, R. Rajamani, L. Alexander, P. J. Starr, and M. Donath, Development and experimental evaluation of a tilt stability control system for narrow commuter vehicles, *IEEE Transactions on Control Systems Technology*, vol. 18, no. 6, pp. 1266–1279, November 2010. `http://ieeexplore.ieee.org/document/5356230/` DOI: 10.1109/tcst.2009.2035819. 8, 10

[24] C. Tang, L. He, and A. Khajepour, Design and analysis of an integrated suspension tilting mechanism for narrow urban vehicles, *Mechanism and Machine Theory*, vol. 120, pp. 225–238, February 2018. `http://www.sciencedirect.com/science/article/pii/S0094114X17306158` DOI: 10.1016/j.mechmachtheory.2017.09.025. 9, 11, 22, 23, 24, 26, 27, 28

[25] J. Berote, J. Darling, and A. Plummer, Lateral dynamics simulations of a three-wheeled tilting vehicle, *Proc. of the Institution of Mechanical Engineers, Part D: Journal of Automobile Engineering*, vol. 229, no. 3, pp. 342–356, February 2015. `https://doi.org/10.1177/0954407014542625` DOI: 10.1177/0954407014542625. 10

[26] R. Rajamani, J. Gohl, L. Alexander, and P. Starr, Dynamics of narrow tilting vehicles, *Mathematical and Computer Modelling of Dynamical Systems*, vol. 9, no. 2, pp. 209–231, June 2003. `http://www.tandfonline.com/doi/abs/10.1076/mcmd.9.2.209.16521` DOI: 10.1076/mcmd.9.2.209.16521. 10

[27] R. Maryniuk, Development of a modular urban electric vehicle, Master's thesis, University of Waterloo, December 2017. `https://uwspace.uwaterloo.ca/handle/10012/12746` 10

[28] M.-A. Rajaie, Design and fabrication of a novel corner wheel module for urban vehicles, Master's thesis, University of Waterloo, September 2016. `https://uwspace.uwaterloo.ca/handle/10012/10929`

[29] A. Waters, A novel universal corner module for urban electric vehicles: Design, prototype, and experiment, Master's thesis, University of Waterloo, August 2017. `https://uwspace.uwaterloo.ca/handle/10012/12217` 10

[30] J. Marsh, Citroën XM technical specification. `http://www.citroenet.org.uk/passenger-cars/psa/xm/xm-09.html` 11

[31] W. Bauer, *Hydropneumatic Suspension Systems*. Springer, 2011. `http://www.springer.com/us/book/9783642151460` DOI: 10.1007/978-3-642-15147-7. 11, 13

[32] C. Tang, Narrow urban vehicles with an integrated suspension tilting system: Design, modeling, and control, Ph.D. dissertation, University of Waterloo, 2018. 12

[33] G. R. Transit, Accessibility on GRT buses, May 2017. http://www.grt.ca/en/rider-information/accessibility-on-grt-buses.aspx 12

[34] T. Motors, Model S owner's manual, December 2018. https://www.tesla.com/sites/default/files/model_s_owners_manual_north_america_en_us.pdf 13

[35] H. B. Pacejka, Tire and vehicle dynamics. *SAE International*, Warrendale, PA, 2006. DOI: 10.1016/C2010-0-68548-8. 16

[36] V. Cossalter, Motorcycle dynamics. Lulu.com, 2006. DOI: 10.1002/9781118536391.ch1.

[37] M. Ataei, C. Tang, A. Khajepour, and S. Jeon, Active camber system for lateral stability improvement of urban vehicles, *Proc. of the Institution of Mechanical Engineers, Part D: Journal of Automobile Engineering*, p. 0954407019832436, March 2019. https://doi.org/10.1177/0954407019832436 DOI: 10.1177/0954407019832436. 16

[38] S. M. Laws, An active camber concept for extreme maneuverability: Mechatronic suspension design, tire modeling, and prototype development, Ph.D. thesis, Stanford University, 2010. 17

[39] C. E. Beal, Applications of model predictive control to vehicle dynamics for active safety and stability, Ph.D. thesis, Stanford University, 2011. 17, 31, 49, 63

[40] M. Ataei, A. Khajepour, and S. Jeon, Reconfigurable integrated stability control for four- and three-wheeled urban vehicles with flexible combinations of actuation systems, *IEEE/ASME Transactions on Mechatronics*, vol. 23, no. 5, pp. 2031–2041, October 2018. DOI: 10.1109/tmech.2018.2862924. 17

[41] R. Rajamani, *Vehicle Dynamics and Control.* Springer, 2012. http://www.springer.com/gp/book/9781461414322 DOI: 10.1007/978-1-4614-1433-9. 18, 30, 33

[42] M. Ataei, A. Khajepour, and S. Jeon, Rollover stabilities of three-wheeled vehicles including road configuration effects, *Proc. of the Institution of Mechanical Engineers, Part D: Journal of Automobile Engineering*, vol. 231, no. 7, pp. 859–871, June 2017. https://doi.org/10.1177/0954407017695007 DOI: 10.1177/0954407017695007. 18

[43] R. Rajamani, *Vehicle Dynamics and Control.* New York, Springer, 2006. DOI: 10.1007/978-1-4614-1433-9. 19

[44] R. Rajamani, D. Piyabongkarn, V. Tsourapas, and J. Y. Lew, Parameter and state estimation in vehicle roll dynamics, *IEEE Transactions on Intelligent Transportation Systems*, vol. 12, no. 4, pp. 1558–1567, December 2011. DOI: 10.1109/tits.2011.2164246. 19

[45] X. Zhang, Y. Yang, K. Guo, J. Lv, and T. Peng, Contour line of load transfer ratio for vehicle rollover prediction, *Vehicle System Dynamics*, vol. 55, no. 11, pp. 1748–1763, November 2017. http://dx.doi.org/10.1080/00423114.2017.1321773 DOI: 10.1080/00423114.2017.1321773. 19

[46] M. Ataei, A. Khajepour, and S. Jeon, A novel reconfigurable integrated vehicle stability control with omni actuation systems, *IEEE Transactions on Vehicular Technology*, vol. PP, no. 99, pp. 1–1, 2017. DOI: 10.1109/tvt.2017.2782569. 21

[47] C. Tang, M. Ataei, and A. Khajepour, A reconfigurable integrated control for narrow tilting vehicles, *IEEE Transactions on Vehicular Technology*, vol. 68, no. 1, January 2019. DOI: 10.1109/tvt.2018.2882382. 21, 33, 38

[48] J. C. Dixon, *Suspension Geometry and Computation*. Wiley, Chichester, UK, December 2009. DOI: 10.1002/9780470682906. 22

[49] T. D. Gillespie, Fundamentals of vehicle dynamics, *Society of Automotive Engineers*, Warrendale, PA, 1992. DOI: 10.4271/r-114. 26

[50] M. Barker, B. Drew, J. Darling, K. A. Edge, and G. W. Owen, Steady-state steering of a tilting three-wheeled vehicle, *Vehicle System Dynamics*, vol. 48, no. 7, pp. 815–830, July 2010. http://dx.doi.org/10.1080/00423110903147474 DOI: 10.1080/00423110903147474. 26

[51] D. J. M. Sampson and D. Cebon, Active roll control of single unit heavy road vehicles, *Vehicle System Dynamics*, vol. 40, no. 4, pp. 229–270, October 2003. http://www.tandfonline.com/doi/abs/10.1076/vesd.40.2.229.16540 DOI: 10.1076/vesd.40.2.229.16540. 30, 31

[52] N. Zhang, G.-M. Dong, and H.-P. Du, Investigation into untripped rollover of light vehicles in the modified fishhook and the sine maneuvers. Part I: Vehicle modelling, roll and yaw instability, *Vehicle System Dynamics*, vol. 46, no. 4, pp. 271–293, April 2008. http://www.tandfonline.com/doi/abs/10.1080/00423110701344752 DOI: 10.1080/00423110701344752. 30, 31

[53] A. Goodarzi, A. Soltani, M. H. Shojaeefard, and A. Khajepour, An integrated vehicle dynamic control strategy for three-wheeled vehicles, *Proc. of the Institution of Mechanical Engineers, Part K: Journal of Multi–body Dynamics*, vol. 229, no. 3, pp. 225–244, September 2015. http://journals.sagepub.com/doi/10.1177/1464419314558741 DOI: 10.1177/1464419314558741. 30

[54] D. Li, S. Du, and F. Yu, Integrated vehicle chassis control based on direct yaw moment, active steering and active stabiliser, *Vehicle System Dynamics*, vol. 46, no. sup1,

pp. 341–351, September 2008. http://www.tandfonline.com/doi/abs/10.1080/00423110801939204 DOI: 10.1080/00423110801939204. 30

[55] Department of Transportation, National highway traffic safety administration. Traffic safety facts 2012: a compilation of motor vehicle crash data from the fatality analysis reporting system and the general estimates system, *National Highway Traffic Safety Administration, Technical Report*, 2012. https://crashstats.nhtsa.dot.gov/Api/Public/ViewPublication/812032 31

[56] J. Yoon, W. Cho, J. Kang, B. Koo, and K. Yi, Design and evaluation of a unified chassis control system for rollover prevention and vehicle stability improvement on a virtual test track, *Control Engineering Practice*, vol. 18, no. 6, pp. 585–597, June 2010. http://www.sciencedirect.com/science/article/pii/S0967066110000729 DOI: 10.1016/j.conengprac.2010.02.012. 31

[57] V. T. Vu, O. Sename, L. Dugard, and P. Gaspar, Enhancing roll stability of heavy vehicle by LQR active anti-roll bar control using electronic servo-valve hydraulic actuators, *Vehicle System Dynamics*, vol. 55, no. 9, pp. 1405–1429, September 2017. http://dx.doi.org/10.1080/00423114.2017.1317822 DOI: 10.1080/00423114.2017.1317822. 31

[58] G.-M. Dong, N. Zhang, and H.-P. Du, Investigation into untripped rollover of light vehicles in the modified fishhook and the sine manoeuvres, part II: Effects of vehicle inertia property, suspension and tyre characteristics, *Vehicle System Dynamics*, vol. 49, no. 6, pp. 949–968, June 2011. https://doi.org/10.1080/00423114.2010.504855 DOI: 10.1080/00423114.2010.504855. 31

[59] S. Yim, Y. Park, and K. Yi, Design of active suspension and electronic stability program for rollover prevention, *International Journal of Automotive Technology*, vol. 11, no. 2, pp. 147–153, April 2010. https://link.springer.com/article/10.1007/s12239-010-0020-6 DOI: 10.1007/s12239-010-0020-6. 31

[60] T. J. Wielenga and M. A. Chace, A study in rollover prevention using anti-rollover braking, *SAE Automotive Dynamics and Stability Conference*, May 2000. http://papers.sae.org/2000-01-1642/ DOI: 10.4271/2000-01-1642. 31, 32

[61] B. Schofield and T. Hagglund, Optimal control allocation in vehicle dynamics control for rollover mitigation, *American Control Conference*, pp. 3231–3236, June 2008. DOI: 10.1109/acc.2008.4586990. 31, 32

[62] J. Yoon and K. Yi, A rollover mitigation control scheme based on rollover index, *American Control Conference*, pp. 5372–5377, June 2006. DOI: 10.1109/acc.2006.1657577. 31

[63] J.-S. Jo, S.-H. You, J. Y. Joeng, K. I. Lee, and K. Yi, Vehicle stability control system for enhancing steerabilty, lateral stability, and roll stability, *International Journal of Automotive*

Technology, vol. 9, no. 5, p. 571, October 2008. https://link.springer.com/article/10.1007/s12239--008-0067-9 DOI: 10.1007/s12239-008-0067-9. 31, 49, 63

[64] J. Gohl, R. Rajamani, L. Alexander, and P. Starr, Active roll mode control implementation on a narrow tilting vehicle, *Vehicle System Dynamics*, vol. 42, no. 5, pp. 347–372, December 2004. http://www.tandfonline.com/doi/abs/10.1080/0042311042000266810 DOI: 10.1080/0042311042000266810. 33

[65] J. W. Robertson, J. Darling, and A. R. Plummer, Combined steering—direct tilt control for the enhancement of narrow tilting vehicle stability, *Proc. of the Institution of Mechanical Engineers, Part D: Journal of Automobile Engineering*, vol. 228, no. 8, pp. 847–862, July 2014. https://doi.org/10.1177/0954407014522445 DOI: 10.1177/0954407014522445.

[66] J. Edelmann and M. Plöchl, Electronic stability control of a narrow tilting vehicle, *SAE International Journal of Materials and Manufacturing*, vol. 4, no. 1, pp. 1006–1013, April 2011. http://papers.sae.org/2011--01-0976/ DOI: 10.4271/2011-01-0976. 33

[67] S.-G. So and D. Karnopp, Active dual mode tilt control for narrow ground vehicles, *Vehicle System Dynamics*, vol. 27, no. 1, pp. 19–36, January 1997. http://dx.doi.org/10.1080/00423119708969321 DOI: 10.1080/00423119708969321. 33

[68] A. Snell, An active roll-moment control strategy for narrow tilting commuter vehicles, *Vehicle System Dynamics*, vol. 29, no. 5, pp. 277–307, May 1998. http://dx.doi.org/10.1080/00423119808969376 DOI: 10.1080/00423119808969376. 33

[69] L. Mourad, F. Claveau, and P. Chevrel, Direct and steering tilt robust control of narrow vehicles, *IEEE Transactions on Intelligent Transportation Systems*, vol. 15, no. 3, pp. 1206–1215, June 2014. http://ieeexplore.ieee.org/document/6718083/ DOI: 10.1109/tits.2013.2295684. 33

[70] C. Tang and A. Khajepour, Integrated stability control for narrow tilting vehicles: An envelope approach, *IEEE Transactions on Intelligent Transportation Systems*, 2019 (under review). 33, 36, 44, 46, 47, 49, 53, 54

[71] T. Haraguchi, T. Kaneko, I. Kageyama, Y. Kuriyagawa, and M. Kobayashi, Study of tilting type personal mobility vehicle by the immersive driving simulator with five large screens, *Transactions of Society of Automotive Engineers of Japan*, vol. 48, no. 3, pp. 693–698, 2017. https://www.jstage.jst.go.jp/article/jsaeronbun/48/3/48_20174447/_article/-char/en 38

[72] M. Jalali, E. Hashemi, A. Khajepour, S.-K. Chen, and B. Litkouhi, Model predictive control of vehicle roll-over with experimental verification, *Control Engineering Practice*, vol. 77,

pp. 95–108, August 2018. https://www.sciencedirect.com/science/article/pii/S0967066118300893 DOI: 10.1016/j.conengprac.2018.04.008. 38

[73] E. Hashemi, Full vehicle state estimation using a holistic corner-based approach, Ph.D. thesis, University of Waterloo, 2017. 43

[74] Y. H. J. Hsu, S. M. Laws, and J. C. Gerdes, Estimation of tire slip angle and friction limits using steering torque, *IEEE Transactions on Control Systems Technology*, vol. 18, no. 4, pp. 896–907, July 2010. DOI: 10.1109/tcst.2009.2031099. 43

[75] A. Rezaeian, R. Zarringhalam, S. Fallah, W. Melek, A. Khajepour, S. Chen, N. Moshchuck, and B. Litkouhi, Novel tire force estimation strategy for real-time implementation on vehicle applications, *IEEE Transactions on Vehicular Technology*, vol. 64, no. 6, pp. 2231–2241, June 2015. DOI: 10.1109/tvt.2014.2345695. 45

[76] E. Hashemi, M. Pirani, A. Khajepour, A. Kasaiezadeh, S.-K. Chen, and B. Litkouhi, Corner-based estimation of tire forces and vehicle velocities robust to road conditions, *Control Engineering Practice*, vol. 61, pp. 28–40, April 2017. http://www.sciencedirect.com/science/article/pii/S0967066117300096 DOI: 10.1016/j.conengprac.2017.01.009. 45

[77] J. B. Hoagg and D. S. Bernstein, Nonminimum-phase zeros—much to do about nothing—classical control—revisited part II, *IEEE Control Systems*, vol. 27, no. 3, pp. 45–57, June 2007. DOI: 10.1109/MCS.2007.365003. 51

[78] L. Qiu and E. J. Davison, Performance limitations of non-minimum phase systems in the servomechanism problem, *Automatica*, vol. 29, no. 2, pp. 337–349, March 1993. http://www.sciencedirect.com/science/article/pii/000510989390127F DOI: 10.1016/0005-1098(93)90127-f. 52

[79] Scokaert Pierre O. M. and Rawlings James B., Feasibility issues in linear model predictive control, *AIChE Journal*, vol. 45, no. 8, pp. 1649–1659, April 2004. https://onlinelibrary.wiley.com/doi/abs/10.1002/aic.690450805 DOI: 10.1002/aic.690450805. 53

[80] M. Brown, J. Funke, S. Erlien, and J. C. Gerdes, Safe driving envelopes for path tracking in autonomous vehicles, *Control Engineering Practice*, vol. 61, pp. 307–316, April 2017. http://www.sciencedirect.com/science/article/pii/S0967066116300831 DOI: 10.1016/j.conengprac.2016.04.013. 53

[81] C. G. Bobier, A phase portrait approach to vehicle stabilization and envelope control, Ph.D. thesis, Stanford University, 2012. 54

[82] S. M. Erlien, S. Fujita, and J. C. Gerdes, Shared steering control using safe envelopes for obstacle avoidance and vehicle stability, *IEEE Transactions on Intelligent Transportation Systems*, vol. 17, no. 2, pp. 441–451, February 2016. DOI: 10.1109/tits.2015.2453404. 54

Authors' Biographies

CHEN TANG

Chen Tang is currently a Postdoctoral Fellow of Mechanical and Mechatronics Engineering with the University of Waterloo, where he earned his Ph.D. in 2018. He received his B.Sc. and M.Sc. in mechanical engineering from Tongji University, China in 2009 and 2012, respectively. His research interests include vehicle dynamics and control, advanced chassis systems, and intelligent transportation systems.

AMIR KHAJEPOUR

Amir Khajepour is a professor in the department of Mechanical and Mechatronics Engineering at the University of Waterloo. He holds the Canada Research Chair in Mechatronic Vehicle Systems, and NSERC/General Motors Industrial Research Chair in Holistic Vehicle Control. His expertise is in several key multidisciplinary areas including system modeling, control, and vehicle dynamics. His research has resulted in many patents and technology transfers. He is the author of more than 500 journal and conference publications as well as several books. He is a Fellow of the Engineering Institute of Canada, the American Society of Mechanical Engineers, and the Canadian Society of Mechanical Engineering.

Printed in the United States
by Baker & Taylor Publisher Services